PACIFIC
SEAWEEDS

PACIFIC SEAWEEDS

A guide to common seaweeds of the West Coast

Louis D. Druehl

Harbour Publishing
P.O. Box 219, Madeira Park, BC Canada V0N 2H0
www.harbourpublishing.com

Harbour Publishing acknowledges the financial support of the Government of Canada through the Book Publishing Industry Development Program (BPIDP) and the Canada Council for the Arts, and the Province of British Columbia through the British Columbia Arts Council, for its publishing activities.
Printed in Canada

THE CANADA COUNCIL | LE CONSEIL DES ARTS
FOR THE ARTS | DU CANADA
SINCE 1957 | DEPUIS 1957

Cover and page design and layout by Martin Nichols
Front cover photo by Russ Norberg
Photograph and illustration credits: AH: drawings from I.A. Abbott and G.J. Hollenberg, *Marine Algae of California* (1976); GWS: Gary Saunders; RKH: Rae Hopkins; TBW: drawing adapted from the notes of T.B. Widdowson; TM: Tom Mumford, Jr.; WRT: drawing adapted from W.R. Taylor, *Marine Algae of the Northeastern Coast of North America* (1962); RFSG: drawing from Robert F. Scagel, *Marine Algae of British Columbia and Northern Washington*, Part 1: Chlorophyceae (National Museum of Canada, 1966), Bulletin 207:1–257. All drawings not otherwise credited are by Ernani G. Menez, from Robert F. Scagel, *Guide to Common Seaweeds of British Columbia* (1967) and *Marine Algae of British Columbia and Northern Washington* (1966). All photographs not otherwise credited are by Russ Norberg.

National Library of Canada Cataloguing in Publication Data

Druehl, Louis D., 1936–
    Pacific seaweeds

Includes bibliographical references and index.
    ISBN 1-55017-240-9

1. Marine algae—Northwest Coast of North America—Identification.
2. Marine algae—Harvesting—Northwest Coast of North America. I. Title.
QK570.5.D78 2001   579.8'817743   C00-911599-4

Dedicated to Professor Robert Francis Scagel,
who made possible my life with kelp,
and to my students, who taught me so much.

CONTENTS

Preface

I am frequently asked how I ended up working with seaweeds. The answer is: through a process of elimination. As a youth I wanted to be a rancher, but I didn't have a ranch so I went to university. I majored in zoology but soon discovered that for me, plants are more intriguing than animals and a lot less messy. I concluded my undergraduate studies by exploring every aspect of botany except seaweeds (I attended an inland school). After university I studied in Europe, and it was there, while rowing on the Adriatic Sea, that I saw my first seaweed. This beckoning seaweed, which I now know to be *Cystoseira*, was a perfectly formed, gently swaying brown bush (a brown bush!). I reached into the sea to touch it and realized it was very large and grew very deep, beyond my reach. And I did then what I have done ever since: I went that extra distance to understand seaweed (except now I keep my clothes on).

Following my Adriatic baptism, I did graduate study, became an academic and discovered the Bamfield Marine Station at Bamfield, BC, where I now live and study surrounded by seaweeds and students. This marine station represents the major marine focus of BC and Alberta universities and attracts students and scientists from all over the world. The most satisfying aspect of my career is that it allows me to explore seaweeds to my heart's content. This guide is my effort to share with you the knowledge I have gained and the joy of discovering Pacific seaweeds.

Acknowledgements

Many people have contributed to this guide. All of the photographs debut here. Russ Norberg (Springfield, Oregon) brought his patience and artistry to the beach and produced the majority of the photographs (all those not credited to others). Photographs of *Porphyra* culture were provided by Tom Mumford, Jr., those of kelp culture, *Costaria* and *Macrocystis* were provided by Rae Hopkins, Gary Saunders provided the photo of *Postelsia* and the Bamfield Marine Station provided the photo *Nereocystis* (#39). This guide is greatly enhanced by its illustrations. Drawings by Ernani G. Menez, which earlier appeared in Professor Robert F. Scagel's (1967) *Guide to Common Seaweeds of British Columbia* and his (1966) National Museum, Ottawa, study of green seaweeds, have been liberally used with Scagel's permission. Several illustrations are from *The Marine Algae of California* by Isabella S. Abbott and George J. Hollenberg, with the permission of the publishers, Stanford University Press, © 1976 by the Board of Trustees of the Leland Stanford Junior University. The drawing of *Fucus spiralis* is adapted from W.R. Taylor's (1962) *Marine Algae of the Northeastern Coast of North America*, the drawing of *Gonimophyllum* is adapted from the notes of T.B. Widdowson, and Anne Stewart provided the drawing of an adult *Macrocystis* plant.

The following correspondents have strengthened my presentations in this guide by their unselfish instruction: Carolyn Bird (National Research Council, Halifax), Carol-Ann Borden (University of BC), Megan N. Dethier (Friday Harbor Laboratories, University of Washington), Robert E. DeWreede (University of BC), Mike S. Foster (Moss Landing Marine Laboratories, Moss Landing, California), Mike S. Graham (University of California, Davis), Gayle I. Hansen (Newport Marine Center, Newport, Oregon), Paul G. Harrison (University of BC), Mimi A.R. Koehl (University of California, Berkeley), Gesile Muller-Parker (Western Washington State University, Bellingham), Tom F. Mumford, Jr. (Department of Natural Resources, Olympia, Washington), Dawn E. Renfrew (Bamfield Marine Station), Leslie Rimmer (Bamfield Marine Station), Gary W. Saunders (University of New Brunswick), Paul C. Silva (University of California, Berkeley), Mike S. Stekoll (University of Alaska), Anne Stewart (Bamfield Marine Station), J. Robert Waaland (University of Washington), and Tom B. Widdowson (University of Victoria).

Mary Schendlinger and the Harbour Publishing crew polished and improved the readability of the guide. My best friend Rae K. Hopkins provided the much-needed spiritual encouragement, served with constructive criticism, to see this project to fruition.

For all of the above, I am thankful.

About Seaweeds

We live on a planet dominated by oceans. As we develop our limited land masses, often converting arable lands into commercial and residential areas and otherwise reducing our ability to support an increasing human population, we will grow more dependent upon ocean resources. To develop these resources wisely, we need to understand the ocean and its inhabitants. Seaweeds are major players in shallow near-shore waters. They constitute the nutritional base for many shallow-water food chains and they provide the architecture to house and protect associated fishes and invertebrates.

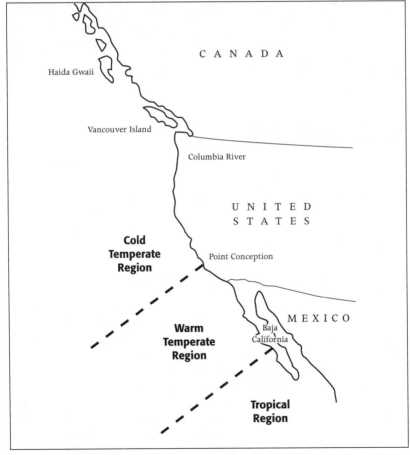

Biogeographic regions of the Pacific coast. This guide covers the Cold Temperate Region.

In addition to playing an important ecological role, seaweeds hold great potential to serve mankind. They are used for food, industrial and scientific chemicals, plant fertilizers and pesticides, and livestock feed supplements, and they are a renewable energy source. Perhaps even more exciting are the potential pharmaceutical uses of seaweeds. Preliminary tests have shown seaweed constituents may combat hypertension (high blood pressure), some cancers, and stroke. Because they are so diverse, and so distinct from flowering plants and fungi, we can expect to discover new, exotic compounds with characteristics beneficial to human health.

The plants dealt with in this guide are major elements of the temperate seaweed flora extending from southeast Alaska (60°N) to central California (34°N, just south of Santa Barbara at Point Conception). This coastline consists of approximately 55,000 km (33,000 miles) of foreshore, most of which is suitable for seaweed habitation. I have not attempted to describe all of the more than 600 species known to inhabit this coast. However, the 134 described species represent the full range of seaweed morphologies and lifestyles in this part of the Pacific Ocean. The common seaweeds you would expect to encounter are included here.

Definition

Seaweeds can be defined as organisms that live in the sea, and that are not all plants but have some basic plant-like features. They are photosynthetic, sedentary and not overtly responsive to external stimuli; they have cell walls that give them rigidity; they are prolific reproducers. Seaweeds do not share other, more advanced features of plants. They do not have flowers, cones or other elaborate enclosed reproductive structures (like moss capsules). They do not have specialized systems for transporting nutrients internally (although some have primitive systems). They do not protect themselves with bark and waxy coverings. Their root-like structures function only to anchor them and are not important in extracting nutrients and water from the soil. They are mostly multicellular and conspicuous. In addition to the approximately 600 seaweed species on our beaches there are a few species of seagrasses, commonly called surf grass or eel grass, which are flowering plants but are included here as honorary seaweeds.

The seaweeds commonly found along the Pacific coast are divided among three groups: the red, green and brown seaweeds. Each group has distinctive pigmentation, storage products and cell wall components. The differences between the groups are great enough that they are considered as not being related, much as we consider a jellyfish and a finfish not to be related. Thus the term "seaweed" encompasses an artificial assemblage of organisms. The features shared by seaweeds mostly define a larger, artificial assemblage, the algae (singular: alga). Seaweeds share the beach with other algae, the blue green algae and diatoms. Most of these are inconspicuous and microscopic, which is not to say inconsequential. The blue green algae are photosynthetic bacteria whose presence is indicated by dark, approaching black, variously shaped little slimy colonies. The diatoms, which are normally encountered floating in the sea (phytoplankton), exist as microscopic epiphytes living on seaweeds, and as dark brown, macroscopic strands attached to rock or seaweeds.

Structure

The cell is the basic compartmental unit of life. It contains those structures necessary for life: nucleus, chloroplast (responsible for photosynthesis), mitochondrion (responsible for the release of chemically bound energy), and others. Among the algae, seaweeds included, there is little cellular differentiation, relative to land plants. The cells of seaweeds may be classified as parenchyma, meaning that they are surrounded by thin cell walls with little or no secondary thickening and they usually retain their ability to undergo cellular division throughout their life. Land plants have parenchyma, collenchyma and schlerenchyma. These latter cell types are characterized by secondary cell wall thickenings, resulting in woody characteristics, and by their inability to divide after reaching maturity.

Seaweed cells are differentiated by the number of nuclei they contain, their association with adjacent cells and their function. The cells may be multinucleate or uninucleate. Generally, cells visible to the naked eye are multinucleate, e.g. *Urospora*, p. 45). Cells are usually separated from adjacent cells by a well-defined end cell wall. During cell division, this cell wall grows as an expanding plate, separating the two daughter cells so that there is limited connection between them. In some seaweeds, the end cell wall is produced when the existing cell wall grows inward, effectively pinching the cell into two daughter cells. There is expanded connection between these cells, which are called coenocytic cells. They are usually multinucleate and may be very long (30 cm/12" in *Codium*, p. 53).

The surface cells of fleshy seaweeds are responsible for photosynthesis and nutrient uptake. The inner cells serve as packing and may form storage and/or primitive conducting tissues. Some cells are responsible for attaching the seaweed to its substrate (no small task). In some seaweeds, such as the green alga *Ulva* (p. 50), almost any cell may become reproductive; but in others, including many red seaweeds, only specific cells become reproductive. Similarly, cells that can undergo frequent division, thus being primarily responsible for growth, may be localized or occur generally throughout the plant.

Seaweeds have three basic growth strategies. In the simplest, termed diffuse growth, cell division is more or less random over the plant body. If you were to mark a grid with India ink or small punched holes on the bladed (leaf-like) green alga *Ulva*, and follow the change in this grid as the plant grew, you would see that the grid expanded but retained the original proportions. This is diffuse growth. In a second type of growth, termed intercalary growth, most cell division is restricted to some location in the plant, but not at the base or tip. In kelp, for example, division takes place between the stipe (stem-like structure) and blade. If you marked a grid on the base of a kelp blade and on the upper portion of the stipe, the grid would expand unevenly over time. The grid points nearest the blade base would be farther apart than the points more distant from the blade base because most of the active cell division takes place within a few centimetres of the base. This is intercalary growth. In the third mode of growth, termed apical growth, most of the growth is associated with one (e.g. *Fucus*, p. 67) or many (most red algae) actively dividing apical cells located at the outer extremes of the plant. A grid marked on a plant with apical growth would expand along the margins of the plant and remain relatively unchanged near the plant centre.

Seaweeds take on a bewildering range of forms. In their extreme, they range from a single-celled green sphere about 1 cm (1/2") in diameter (*Derbesia*, p. 48) to what appears to be a large (up to 36 m long) brown onion (*Nereocystis*, p. 90). Seaweeds may be filamentous, branched or not. They may be fleshy crusts, blades, tubes or bushes, flattened or radially branched. All of these forms are represented in the red, green and brown seaweed groups. This repetition of form has led Diane and Mark Littler (Smithsonian Institution, Washington, DC), Steve Murray (California State University, Fullerton), and various colleagues to consider the functional role of form. Does a particular form convey a competitive advantage? Are some forms more resistant to grazing? Do different forms dominate during different stages of succession (the development of a mature community from virgin substrate)? These and related questions are discussed in Identifying Pacific Seaweeds (pp. 24–126) and Seaweed Ecology (pp. 127–36).

Reproduction

The life cycle of humans, like that of almost all animals, is simple and straightforward. Your cells are diploid (having two sets of chromosomes). Some of these cells will undergo meiosis, a type of cell division that reduces the chromosomes to one set, producing haploid (one set

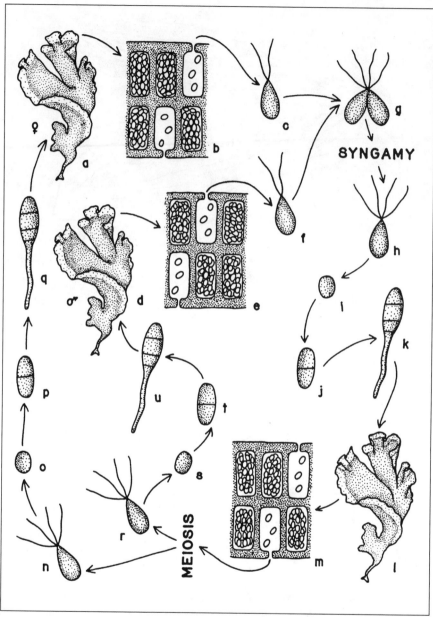

The isomorphic life cycle of *Ulva* (sea lettuce). The haploid sexual plants, gametophytes (a, b and d, e), produce morphologically similar gametes (c, f), which sexually fuse (g, h). The resulting zygote (fertilized gamete) (i) develops into the diploid spore-producing phase, the sporophyte, (j–m), which produces spores by meiosis (reduction cell division) (n, r). These spores develop into the new gametophyte generation (o–q, s–u).

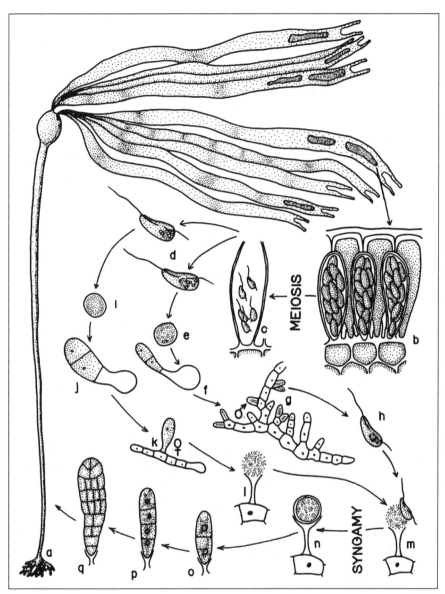

The heteromorphic life cycle of *Nereocystis* (bull kelp). The macroscopic spore-producing phase, the sporophyte (a), produces spores in well-defined patches (sori) by meiosis (reduction cell division) (b, c). These spores are released from the sori after the sori have been dropped (c, d), and they develop into microscopic male (e–g) and female (i–k) gameto-phytes. The sperm (h) is attracted to the retained egg (l) and fertilization occurs (m). The resultant zygote (fertilized egg cell) (n) develops into the new sporophyte generation while attached to the female gametophyte.

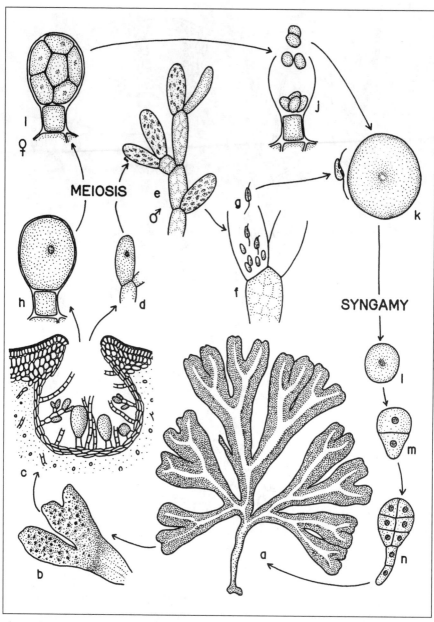

The animal-like life cycle of *Fucus* (rock weed). The macroscopic plant (a) develops special cells that produce sperm (d) and eggs (h) in pits (conceptacles) (c) located on its branch tips. Meiosis (reduction cell division) occurs in these special cells, producing eggs (i–k) and sperm (e–g). The sperm is attracted to the released egg, with which it fuses (k). The resultant zygote (l) develops into the macroscopic sexual plant (m, n).

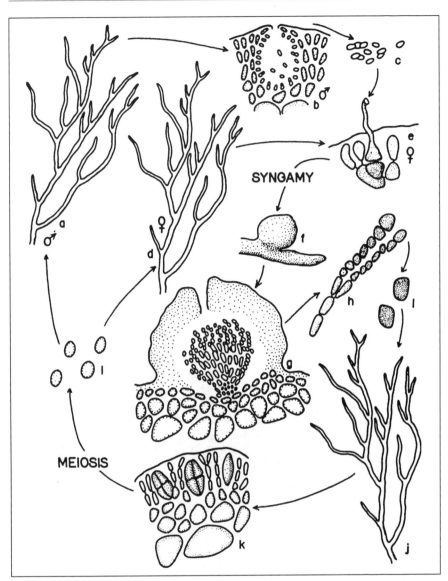

The three-phased life cycle of the red seaweed *Gracilariopsis*. Separate haploid gametophytes, male (a) and female (d), produce non-motile sperm (b, c) and attached eggs (e). The sperm encounters a bowling pin-shaped egg and fuses with it (e). The resulting zygote develops into a parasitic diploid phase on the female gametophyte (f, g). This parasitic phase produces spores (h, i), which are released and develop into a free-living diploid phase (j). The free-living phase then produces spores (k, l) by meiosis (reduction cell division). These spores develop into male and female gametophytes—the sexual phase.

of chromosomes) eggs or sperm. The fusion of the egg and sperm introduces two chromosome sets into a zygote. The zygote increases in cell number by mitosis, a type of cell division that does not change the number of chromosome sets, and develops into a mature diploid individual.

The life cycle of plants, particularly the seaweeds, is much more elaborate and varied than that found in animals. In plants, the products of meiosis are rarely eggs and sperm but rather a haploid plant body. Thus a plant can have two bodies, a haploid one and a diploid one. The diploid body is called the sporophyte because it produces spores by meiosis. These spores give rise to the haploid body, which is called the gametophyte because it gives rise to gametes (eggs, sperm or other sexually active cells).

The gametophytes and sporophytes may appear very similar, a condition referred to as isomorphic (same form) or they may be dissimilar, or heteromorphic (different form). The green seaweed *Ulva* (sea lettuce, p. 50) has a life cycle that alternates between morphologically similar gametophytes and sporophytes (isomorphic) (see figures on pp. 15–18. Some seaweeds have gametophyte and sporophyte generations that are markedly different in appearance (heteromorphic). In the brown seaweed *Nereocystis* (bull kelp, p. 90), the sporophyte is a very large plant—often longer than 30 m—and the gametophyte is a small filamentous plant, not visible to the naked eye. This reduction of the gametophyte, relative to the sporophyte, has its greatest expression in the brown seaweed *Fucus* (p. 67), where the gametophyte is reduced to eggs and sperm as in the typical animal life cycle. At the other extreme are green seaweeds such as the green filamentous *Urospora* (p. 45), which has a single-celled sporophyte and a multicellular filamentous gametophyte.

The life cycles of most red seaweeds have an added twist. The gametophytes release their sperm but retain their eggs. The eggs are fertilized while attached to the gametophyte and the new sporophyte generation develops parasitically on the gametophyte. This sporophyte produces spores by mitosis (no reduction in chromosome sets). These spores are released and establish a free-living sporophyte generation. The sporophyte undergoes meiosis (reduces the chromosome sets to one) in the production of spores that will give rise to a new gametophyte generation. As with other seaweeds, the red seaweeds may have isomorphic or heteromorphic generations.

The various types of life cycle may convey survival advantages. For example, some red and brown seaweeds have two morphologies, a persistent crust and a short-lived erect bladed (leaf-like) structure. The crust may be less susceptible to grazing or summer drying than the more delicate blade.

Many seaweeds are capable of reproducing outside of their life cycle. This type of reproduction is referred to as asexual reproduction.

The simplest form of asexual reproduction is fragmentation, where broken-off bits of plant develop into new individuals. A more sophisticated form of asexual reproduction is spore production. Spores, produced by mitosis, are released, often by the hundreds, and develop into new individuals. All progeny resulting from asexual reproduction are genetically identical (clones). Asexual reproduction is a means of increasing an individual's dominance in its environment.

Names

Every known species has a scientific name, and some species have common names. The scientific name is unambiguous as it applies to only one particular species and it is accepted by scientists all over the world. This name is usually derived from Latin or Greek. It may convey a description (e.g. *Enteromorpha*=intestine form) or not (e.g. *Lola*=a friend's name?). The common name, the name by which a species becomes known among lay people, may relate to any of several species and often is haphazardly applied. Some common names are universally applied and are very useful, as with breeds of dogs and species of birds. A few seaweeds have well-established common names, such as *Nereocystis luetkeana* (bull kelp) and *Codium fragile* (stag horn). In this guide, common names—some of them well established, some not—are included. But the majority of our seaweeds do not have widely accepted common names. My advice is to use a scientific name and attach a descriptive phrase: *Costaria costata*, the five-ribbed kelp.

Species of seaweed are classified in a taxonomic hierarchy that reflects their relatedness. This hierarchy is universally applied to all living things and some that are not (fossils). For example, *Laminaria saccharina* (sugar kelp), a brown seaweed, is classified as follows:

Kingdom: Protista. A kingdom is a large grouping of vaguely similar organisms. Other kingdoms include Animalia and Plantae.

Division: Phaeophyta (brown algae). A division is a group of organisms thought to share a common ancestor. The animal equivalent to division is phylum. The -phyta suffix designates divisional status. Other divisions include Chlorophyta (green algae) and Rhodophyta (red algae).

Class: Phaeophyceae. Classes designate groups within a division that have significant differences but that are closely related. For example, birds and mammals form separate classes within the phylum Chordata (animals with backbones). The -phyceae suffix designates class. Other classes include Chlorophyceae (most green algae) and Bacillariophyceae (the diatoms).

Order: Laminariales. A class is subdivided into orders on the basis of features such as body plan and life cycle. In the brown seaweeds, the order Laminariales (large brown algae, the kelp) have intercalary growth, the Ectocarpales (small filamentous brown algae) have diffuse growth and the Fucales (common brown rock weeds) have apical growth. The -ales suffix designates ordinal status.

Family: Laminariaceae. Orders are subdivided into families on the basis of various features. Most members of the laminarialean family Alariaceae (the winged kelp) have sporophylls (special blades for spore production) and no branching. Most members of the laminarialean family Lessoniaceae are regularly branched and do not have sporophylls. The Laminariaceae lack sporophylls and branching. The -aceae suffix designates familial status.

Genus: *Laminaria*. When I was defending my PhD thesis I was asked, "What is a genus?" I went blank (aspiring biologists are always asked, "What is a species"!). We can define a genus (plural: genera) as a group of closely related species that are usually not interfertile and are distinguished on the basis of some relatively small morphological feature, such as branching pattern. There may be several genera in a family. The generic name is always italicized or underlined.

Species: *Laminaria saccharina* (L.) Lamouroux. Individuals making up a species are thought to be sexually compatible. Different species within a genus are considered to be sexually incompatible or at least sexually isolated. In reality, this sexual criterion is rarely confirmed. Most species recognized today were established on the basis of morphological features prior to our understanding of the significance of interbreeding and sexual isolation. The species name is a binomial (two names), including the generic name, and it is italicized or underlined. Following the species name are other names (not italicized), which give more information on the species. The "(L.)" in the species name above tells us that Linnaeus was the first to describe this seaweed. He called it *Fucus saccharina* L. Later, Lamouroux established this seaweed in the genus *Laminaria*. Occasionally I see an old reference to *Fucus saccharina* L., and I can track it to its modern name. "Species" is both singular and plural. ("Specie," as in *in specie*, is to pay in coin.)

Beach Etiquette, Collections, Safety

As you stand on the shore, enjoying the splendour around you, consider the myriad plants and animals gasping under your boots. Anyone who ventures into the intertidal region has an adverse impact on its inhabitants, but we can minimize this impact. If you turn a rock to see what is under it, roll it back into its original position. Collect specimens only when you must. The seaweeds described in this guide can be identified in the field, without being disturbed. British Columbia and all western US states have regulations governing the collection of seaweeds; become aware of these. The Nutrition and Cooking section (p. 153) points out the advantages and joys of eating seaweed foods, but this is not an endorsement of wild foraging. Sea vegetables are commercially available, having been produced on farms or harvested from the wild under controlled, sustainable conditions.

If you must make collections, do so sparingly and never remove a rare plant—take a photograph. A correct, scientific collection requires that the entire plant be removed, as the nature of the attachment organ along with the more conspicuous plant parts are important in making detailed identifications. If you collect into plastic bags, remember that they turn into miniature greenhouses if exposed to the sun, quickly cooking your algae. Keep specimens of the brown seaweed *Desmarestia* (p. 75) separate from other seaweeds. *Desmarestia* releases sulfuric acid when in captivity, which will destroy your other plants.

The best way to preserve seaweeds attractively is to press them to dryness. Arrange fresh plants on quality paper such as herbarium paper. Fine-structured plants may be floated on the paper from a shallow tray containing sea water and then teased into an attractive arrangement with a watercolour brush and needle. Then cover the plants with a clean cotton cloth and press them in a plant press. A plant press is an elaborate sandwich: outer layers of corrugated cardboard (which allow for the passage of air through the sandwich), layers of absorptive paper (newspaper will do) inside the cardboard (to draw moisture away from the plant to the cardboard), and the herbarium paper, with its arranged plant, covered with a clean cloth. This process may be repeated over and over, creating a stack of pressed plants. The stack is sandwiched in with plywood and compressed with weights or cinched straps. When the plants are dry (they no longer feel cool to the touch), the cloth is carefully peeled off to reveal an attractive plant adhering to the paper by its own glue—a glue that prefers paper to cloth. For thick plants, it may be necessary to replace the newspaper to dry the plant fully. This process will produce attractive preserved specimens of many seaweeds, a notable exception being the coralline algae (red seaweeds with a hard calcareous covering, p. 97).

> Fair spread on pages white, I saw arrayed
> These fairy children of a sire so stern;
> Their beauty charmed me.
> —Appleton.

Whenever you are exploring the seashore, remember that the beach is a dangerous, adrenaline-jerking place. Mark Denny (Stanford University), who explored adaptations of west coast intertidal organisms to prevalent drag forces, projected the force generated by the largest expected wave for any given year as being equivalent to 13 tonnes (14 tons) on the average human form. This illustrates the challenges intertidal organisms face and a good reason to be very cautious while studying in a wave-exposed environment. Always have someone watch for menacing waves. Professor Gilbert M. Smith (Stanford University), author of the pioneer study *Marine Algae of the Monterey Peninsula, California* (1944), advised that if you get caught by a wave, do not run—lie down and act like a sea star.

Identifying Pacific Seaweeds

To identify a seaweed species with this guide:

1 Flip to the Thumbnail Identification Guide at the back of the book and read down the list to narrow your search.

2 Go to the page reference noted in the Thumbnail Guide for more details on the species.

3 If a photo reference is shown for the species, go to the Colour Guide and compare your specimen with the photograph.

Green Seaweeds and Seagrasses

Seagrasses

Zostera (Eel grass, p. 42)

1 *Zostera marina* (below left) with *Lola lubrica* (above).

2 *Melobesia mediocris* on *Zostera marina*.

Phyllospadix (Surf grass, p. 43)

3 *Phyllospadix scouleri* with seed pods.

Filaments

Unbranched filaments

Lola (Green fish line, p. 45)

4 *Lola lubrica* (upper right) tangled in *Zostera marina*.

Branched filaments

Acrosiphonia
(Green rope, p. 46)

5 *Acrosiphonia coalita*.

Derbesia (Green sea grape– filamentous stage, p. 48)

6 *Derbesia marina* (filamentous phase) in a bear print.

7 *Derbesia marina* (spherical phase) on *Pseudolithophyllum muricatum.*

8 *Derbesia* (spherical phase) on *Lithothamnion phymatodeum* on limpet.

Blades

Prasiola
(Short sea lettuce, p. 49)

9 *Prasiola meridonalis.*

Ulva (Sea lettuce, p. 50)

10 *Ulva fenestrata.*

Cylinders

Enteromorpha
(Green string lettuce, p. 52)

11 *Enteromorpha intestinalis* submerged.

Spongy texture

Codium (p. 53)

12 *Codium fragile* (Sea staghorn) with *Hedophyllum sessile*.

13 *Codium setchellii* (Green spongy cushion) with *Bossiella* sp.

Unicellular Forms (p. 55)

14 *Chlorella*-like plants inhabiting the tissue of *Anthopleura* (sea anemone).

Brown Seaweeds

Crusts

Ralfsia (Sea fungus, p. 58)

15 *Ralfsia fungiformis.*

Globular or cushiony

Colpomenia
(Oyster thief, p. 62)

16 *Colpomenia peregrina.*

Leathesia
(Sea cauliflower, p. 63)

17 *Leathesia difformis.*

Small flat blades

Punctaria (Brown sieve, p. 64)

18 *Punctaria* sp. on *Zostera marina.*

Phaeostrophion
(Sand-scoured false kelp, p. 64)

19 *Phaeostrophion irregulare* in sand.

Petalonia (False kelp, p. 65)

20 *Petalonia fascia* with *Enteromorpha* sp.

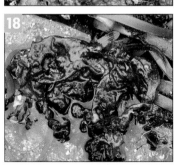

Dichotomously branched

Fucus (Rock weed, p. 67)

21 *Fucus gardneri.*

Pelvetiopsis
(Dwarf rock weed, p. 68)

22 *Pelvetiopsis limitata.*

Cylinders

Scytosiphon
(Soda straws, p. 70)

23 *Scytosiphon lomentaria* with *Acrosiphonia* sp.

Analipus
(Fir branch—erect phase, p. 72)

24 *Analipus japonicus* (erect phase).

Desmarestia (Stringy acid hair—cylindrical form, p. 72)

25 *Desmarestia aculeata.*

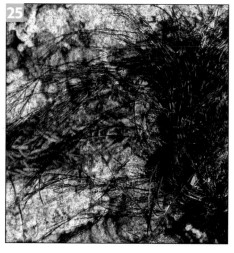

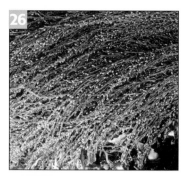

Radially branched with some parts flattened

Sargassum
(Japanese weed, p. 74)

26 *Sargassum muticum.*

Feather-like branching

Desmarestia
(Flattened acid leaf, p. 75)

27 *Desmarestia* sp. (note midribs) with *Macrocystis integrifolia.*

Kelp
Simple blades

Laminaria (Tangle, p. 78)

28 *Laminaria setchellii.*

29 *Laminaria saccharina.*

Hedophyllum
(Sea cabbage, p. 81)

30 *Hedophyllum sessile.*

Winged kelp

Alaria (Winged kelp, p. 82)

31 *Alaria marginata.*

32 *Alaria nana.*

Pterygophora
(Walking kelp, p. 84)

33 *Pterygophora californica* with geniculate coralline.

Egregia (Feather boa, p. 85)

34 *Egregia menziesii.*

35 *Egregia menziesii* bed smothering *Laminaria setchellii.*

Eisenia (Forked kelp, p. 86)

36 *Eisenia arborea.*

Ribbed kelp

Pleurophycus
(Broad rib kelp, p. 87)

37 *Pleurophycus gardneri.*

Costaria (Five rib kelp, p. 88)

38 *Costaria costata*, wave-sheltered (wide) and wave-exposed (narrow) forms. RKH

Repeatedly branched kelp

Nereocystis (Bull kelp, p. 90)

39 *Nereocystis luetkeana.* BMS

Postelsia (Sea palm, p. 91)

40 *Postelsia palmaeformis.* GWS

Lessoniopsis
(Strap kelp, p. 92)

41 *Lessoniopsis littoralis.*

Macrocystis (Giant kelp, p. 93)

42 *Macrocystis integrifolia.* RKH

43 *Macrocystis integrifolia* bed. RKH

Red Seaweeds

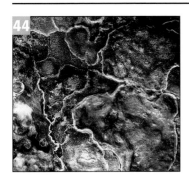

Hard crusts

Pseudolithophyllum
(Rock crust, p. 96)

44 *Pseudolithophyllum muricatum.*

45 *Lithothamnion* sp. on limpet with *Derbesia* (spherical phase).

Melobesia
(Seagrass crust, p. 96)

46 *Melobesia mediocris* on *Zostera marina*.

Mesophyllum
(Coralline crust, p. 97)

47 *Mesophyllum* sp. on geniculate coralline.

Branched and hardened

Corallina (Coral seaweed, p. 98)

48 *Corallina officinalis*.

49 *Corallina vancouveriensis*.

Bossiella (Coral leaf, p. 99)

50 *Bossiella* sp.

51 *Mesophyllum* sp. on *Bossiella* sp.

Calliarthron (Bead coral, p. 99)

52 *Calliarthron* sp.

Soft crusts

Mastocarpus (Turkish washcloth—crust phase, p. 101)

53 *Mastocarpus papillatus* (crust and blade phases).

54 *Mastocarpus papillatus* (blade and crust phases) (below) with *Porphyra* (above).

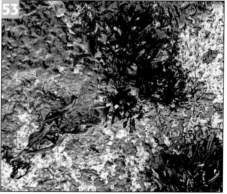

Parasitic plants

Gonimophyllum (p. 102)

55 *Gonimophyllum skottsbergii* parasitic on *Cryptopleura*.

Filamentous

Callithamnion
(Beauty bush, p. 106)

56 *Callithamnion pikeanum*.

Ceramium (Staghorn felt, p. 106)

57 *Ceramium* sp.

Solid or hollow cylinders

Halosaccion
(Dead man's fingers, p. 108)

58 *Halosaccion glandiforme*.

Nemalion
(Rubber threads, p. 109)

59 *Nemalion helminthoides.*

Blades with regular bumps

Chondracanthus
(Turkish towel, p. 110)

60 *Chondracanthus sp.* with *Macrocystis integrifolia* (below) and *Codium fragile* (above).

Mastocarpus (Turkish washcloth–erect phase, p. 110)

61 *Mastocarpus papillatus* (crust and bladed phases) with *Fucus* (above left) and *Halosaccion* (above).

62 *Mastocarpus papillatus,* blade (left) and crust phases.

Blades with ribs

Cryptopleura
(Hidden rib, p. 112)

63 *Cryptopleura* with parasitic *Gonimophyllum skottsbergii.*

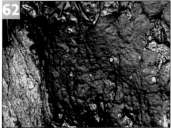

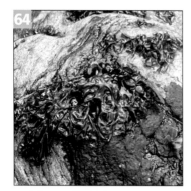

Simple blades

Porphyra (Purple laver, p. 114)
64 *Porphyra* sp.

Palmaria (Red ribbon, p. 115)
65 *Palmaria mollis.*

Mazzaella
(Rainbow leaf, p. 116)
66 *Mazzaella cornucopiae.*
67 *Mazzaella splendens.*

Constantinea
(Cup and saucer, p. 117)
68 *Constantinea simplex* surrounded by geniculate corallines.

Branched in one plane, flattened

Prionitis (Bleach weed, p. 118)

69 *Prionitis*, bleached.

Osmundea
(Red sea fern, p. 119)

70 *Osmundea spectabilis.*

Microcladia (Sea lace, p. 120)

71 *Microcladia coulteri* pressed plant.

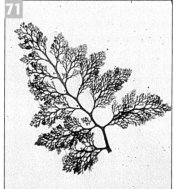

Bushy branched

Gelidium (Gel weed, p. 123)

72 *Gelidium.*

Odonthalia (Sea brush, p. 123)

73 *Odonthalia* pressed plant.

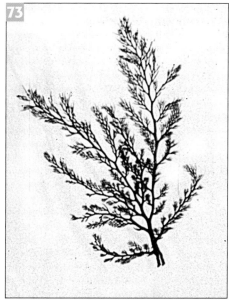

74 *Odonthalia floccosa.*

Neorhodomela
(Black larch, p. 124)

75 *Neorhodomela larix.*

Endocladia (Sea moss, p. 125)

76 *Endocladia muricata.*

Gastroclonium
(Sea belly, p. 125)

77 *Gastroclonium subarticulatum.*

Gracilaria
(Red spaghetti, p. 126)

78 *Gracilaria* sp.

Descriptions and Notes for Pacific Seaweeds

I have not provided specific information on distributions of the species included, because most known ranges in distribution probably understate the real distribution, and species are often discovered where they have not been seen previously. To assist you in remembering scientific names, I have included my translations of the Latin and Greek names, although some of these translations would make a scholar of classics squirm. Also, when available, common names are provided.

Derbesia, the green sea grape co-inhabits a red hard crust with a limpet and a chiton.

Green Seaweeds and Seagrasses

Green seaweeds are part of a larger group, the green algae (Division Chlorophyta), which inhabit almost every conceivable habitat—soil, snow, air, inside rock and polar bear hair. There are approximately 8,000 green algal species on Earth, most of which are associated with fresh water. The greens are distinct from the brown and red seaweeds by virtue of their pigmentation, the chemical nature of their storage products and cell walls, and ultra structural details such as chloroplast (cell structure responsible for photosynthesis) and flagella (whip-like appendages for movement). All green plants have similar pigments, with chlorophyll being the dominant form, which accounts for their green colour. The major storage product of most green plants is starch.

The marine green seaweeds of our area are represented by swimming and non-swimming unicellular forms, filaments, sheets, and globular and cushion-like forms. Close to 130 green seaweed species are recognized locally. Six basic forms (including seagrasses) are described here; they represent the diversity of species found from Alaska to central California. Most of these forms are distributed through southern California into Mexico as well.

Seagrasses

Seagrasses are flowering plants that have adapted to the marine environment. These plants form ecologically and energetically important meadows in the lower intertidal and upper subtidal regions. Studies by Paul G. Harrison (University of British Columbia), Peter McRoy (University of Alaska) and Ron Phillips (Seattle University) have defined the importance of these plants to animal populations inhabiting and visiting these meadows. For example, seagrasses have a level of productivity approaching that of an intensely managed Puerto Rico alfalfa field (see Productivity, p. 136).

Zostera
Eel grass

Photos 1 & 2, p. 25

Zostera (Greek=aquatic plant) is represented by two species, one of which, *Z. marina*, is common. *Zostera marina* has dull green leaves usually wider than 4 mm (3/16"), flattened in cross-section, and up to 3 m (10') long. Plants are usually in wave-protected areas, rooted in mud or muddy sand from Alaska to Mexico. *Zostera japonica* is an uncommon eel grass that was accidentally introduced from Japan. It may have made its way to our shores as packing for Japanese oyster spat (see *Sargassum*, p. 74).

Zostera marina and other seagrasses have considerable potential in habitat conservation. Unlike seaweeds, seagrasses are adapted to living on soft marine bottoms, so they may be planted on unstable or disturbed soft bottoms, where they can curb erosion by holding down and trapping sediment. This is not to say that seagrass beds themselves are immune to erosion. Changes in local current patterns or scarring of the beds by vehicles, etc., may destabilize the sediments, bringing about destruction of the grasses and all that depend upon them.

The leaves of *Zostera* were used as mattress stuffing and

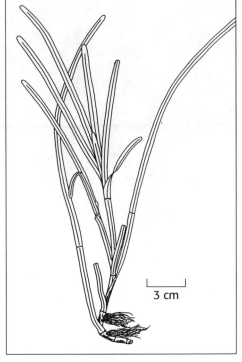

Zostera marina.

house insulation earlier this century. The loss of eel grass beds in the north Atlantic Ocean, which was associated with the population explosion of a marine slime mould, brought about the collapse of the Nova Scotia eel grass mattress stuffing industry and severely reduced populations of associated marine animals. Subsequently, synthetic materials have replaced eel grasses in these industrial uses.

Phyllospadix
Surf grass
Photo 3, p. 25

Phyllospadix (Greek=leaf and spike of flowers) is represented by three species. The blades are bright green, usually less than 4 mm (3/16") wide. The plants are rooted (attached) to solid rock. A covering of sediments may obscure this attachment. Paul G. Harrison has observed surf grass roots buried deeper than 50 cm (20") by accumulated sediments. *Phyllospadix scouleri* and *P. serrulatus* have blades 2–4 mm (1/16–3/16") wide and usually less than 1 m (3') long. These species grow in wave-exposed areas in the lower intertidal and subtidal regions, and tide pools. They may be distinguished by the presence of transparent

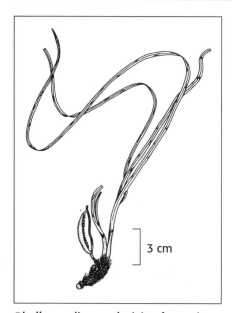

3 cm

Phyllospadix scouleri (surf grass) with seed pod.

teeth along the margin of young *P. serrulatus* leaves, which can be seen using a magnifying glass. The third species, *P. torreyi*, has blades 2 mm (1/16") wide or narrower, and up to 3 m (10') long. This species is found intertidally and subtidally, below *P. scouleri*. *Phyllospadix torreyi* is frequently found in channels and back eddies, where sand accumulates. *Phyllospadix* is distributed from Alaska to Mexico.

As flowering plants have adapted their reproduction to the marine environment, some curious morphological features have evolved. Seagrass pollen may occur in a variety of strange shapes: thread-like, spherical, boomerang-shaped. Joe Ackerman (University of Northern BC) has studied pollination in seagrasses having filamentous (thread-like) pollen. This pollen tumbles through the water, sweeping large areas, thus enhancing its chance of encountering the stigma (the pollen-receptive part of the female flower). Following fertilization, the seagrass seed and surrounding seed fruit develop and then are released. Hooks on the fruit/seed combo are adapted to catch more effectively on particular seaweed morphologies. Once snared, the seed germinates and sends down roots to secure the plant to the substrate.

Filaments

Filamentous (thread-like) green seaweeds are characterized as being one cell wide and variously branched or unbranched. Some forms may remind the insensitive explorer of scum. Of all the seaweeds, these plants are the most similar to most macroscopic freshwater algae. Their general appearance suggests a tight-knit assemblage, but microscopic features distinguish the 35-plus local species into a wide range of diverse green algal groups. They may have one or several nuclei per cell, the filament may be compartmentalized by cross-walls or not (making them essentially unicellular) and they have many types of life cycles. One life cycle incorporates both a filamentous and globular form (see *Derbesia*, p. 48). The following examples express the morphological range of common green filaments.

Unbranched filaments

Lola
Green fish line
Photo 4, p. 25

Lola is represented by one species, *L. lubrica.* This suggestively named seaweed is an unbranched and usually unattached filament. Its multinucleate cells are invisible to the naked eye and its overall look and feel are reminiscent of green fishing line. Plants are usually in excess of 20 cm (8") long. They are often found in wave-sheltered waters, entangled with other attached seaweeds in the mid- to lower intertidal region, from BC to central California.

Ulothrix
Mermaid's tresses

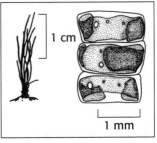

Ulothrix (Greek=shaggy hair) is represented locally by six species. Their unbranched filaments are composed of microscopic cells, each having one nucleus. The chloroplast has the unique shape of an incomplete band, like a cheap engagement ring. The filaments (usually less than 10 cm (4") long) are attached to wood or rock high up in the intertidal zone. This group is represented from Alaska to central California. They are one of very few seaweed genera inhabiting both marine and freshwater areas.

Ulothrix flacca. Drawing of cellular arrangement (right) shows ring-shaped chloroplast.

Urospora
Hanic's green barrels

Urospora (Greek=tailed spore, named after a microscopic stage in the life cycle), represented locally by four species, is an attached, unbranched filament composed of large, barrel-shaped cells, visible to the naked eye (but a magnifying glass helps). The filaments may reach 30 cm (12") in length but usually are less than 10 cm (4") long. *Urospora* grows on rock and wood in the mid- to high intertidal region from Alaska to southern California. In many ways it is very similar to *Ulothrix* but may be distinguished by its larger cells and its lower position in the intertidal region.

Locally, Louis Hanic (recently of the University of Prince Edward Island) was able to link *Codiolum*, a microscopic bowling pin-shaped

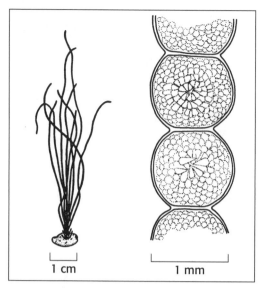

Urospora penicilliformis. At right is detail of cell shape.

unicellular green alga that lives inside fleshy red algae, to the life cycle of *Urospora* (see unicellular forms, p. 55). Previously, *Codiolum* was classified in a distant and distinct group of green algae. This discovery was the culmination of four years of graduate research (University of BC, 1965) by a former antique furniture restorer from Secovce, Czechoslovakia.

Branched filaments

Acrosiphonia
Green rope

Photo 5, p. 25

Acrosiphonia (Greek=tube-topped) is locally represented by six species. Typically, the attached and branched plants form rope-like strands that result from hooked branches tangling the thread-like filaments together. Often these hooked filaments can be seen with the aid of a magnifying glass. More conspicuous forms may reach 40 cm (16") in length. This group of multinucleate plants (many nuclei per cell) has been associated with a *Codiolum*-like phase (see *Urospora*, above). Plants are often abundant in the mid- and low intertidal regions, distributed from Alaska to central California.

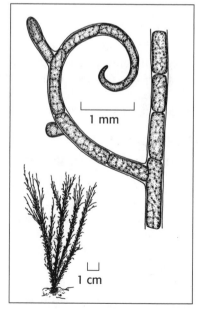

Acrosiphonia coalita and detail (right) showing characteristic hooks.

Cladophora
Green tuft

Cladophora (Greek=branch bearer) is represented locally by ten species. It is an example of a complex assemblage of branched filamentous seaweeds including *Rhizoclonium* (two species) and *Chaetomorpha* (ten species). Plants are typically densely branched but lack the hooks of *Acrosiphonia* (above), and therefore form loose clumps or extensive mats, which may reach 30 cm (12") in length but usually are less than 5 cm (2") long. The cells are microscopic and multinucleate. Plants are usually attached to rock, occupying high tide pools as well as lower exposed surfaces. *Cladophora* is distributed from Alaska to Mexico.

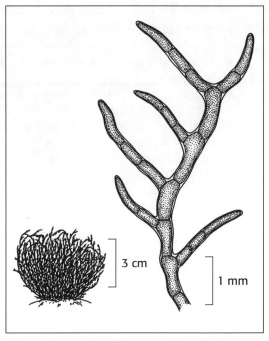

Cladophora columbiana and detail of filament arrangement (right).

Cladophora aegagropila, a freshwater species, forms filamentous balls that may reach 30 cm (12") in diameter. The balls gracefully and repeatedly move from lake bottom to lake surface: at sunrise the balls are lifted off the lake bottom by oxygen bubbles produced by photosynthesis and trapped by the filamentous ball. When the ball reaches the surface, some of the oxygen is released to the atmosphere and the ball descends until enough oxygen is produced to lift it again. In Japan, where the species is known as Marimo, people were so captivated by this activity that they imprisoned virtually all the wild *Cladophora* balls in their private pools. Emperor Hirohito recognized the danger to the species and, upon releasing his *Cladophora* balls, decreed that the species be protected. A Japanese postage stamp commemorates this event, and *C. aegagropila* has the distinction of being the only species protected by imperial decree.

Derbesia
Green sea grape (filamentous stage)
Photos 6, 7, 8, p. 26

Derbesia (after Alphonse Derbes, a French phycologist) is locally represented by *D. marina*, which is commonly distributed from Alaska to southern California. This species has a life cycle that consists of two different morphologies: an inconspicuous branched filamentous phase lacking normal cross-walls, alternating with a unicellular spherical (up to 1 cm/$1/2$" in diameter) phase. The filamentous plants form fuzzy tufts with branches up to 1.5 cm ($3/4$") long. Both phases may be found on rock or crustose coralline seaweeds in the low intertidal and subtidal zones of wave-swept shores. The filamentous phase may also be found on sand.

The spherical phase produces sexually fusing gametes fortnightly. These gametes fuse, initiating the spore-producing filamentous stage. Spheres with dark patches produce female-like swimming gametes and spheres with pale, yellowish patches produce male-like swimming gametes. These patches explode, deflating the spherical cell and releasing the gametes. The gametes fuse in the surf zone (no small feat!), settle out and develop into the filamentous phase. Over the following two weeks the sphere is repaired and re-inflated, and more gametes are produced.

1 mm

3 cm

Bryopsis
Green sea fern

Bryopsis (Greek=moss-like) is represented by three species. The typical feather-like branched filament lacks normal cross-walls and is essentially unicellular. Thus a typical plant represents one cell up to 15 cm (6") long and is branched. Plants are found on rocks, shells and wood in the lower intertidal region from BC to Mexico.

These giant cells could "bleed" to death if they broke. *Bryopsis* overcomes this problem by taking protein bodies from throughout the cell and moving them quickly to the point of break in

Bryopsis corticulans and detail of a large cell (left).

order to seal the rupture. Scientists have exploited these giant cells by conducting unique hybridization experiments. The procedure is to squeeze the protoplasm from each of two different morphological forms and to mix them together. The resulting plant (a hybrid) contains nuclei representing the different morphologies in the same cell. The form taken by this engineered plant will reveal which morphology is genetically dominant or if neither is dominant, which would result in an intermediate morphology.

Blades

The blade morphology is common among the major seaweed groups. In the green seaweeds, the blades are usually one or two cells thick. The blade's thickness as well as its other dimensions are determined by the pattern of cell division. A game biologists play is to arrange species on the basis of morphological criteria. For example, one could establish a simple-to-complex lineage for some of the common green seaweeds. Imagine the green seaweed cell as being a cube. In an unbranched filament, this cube can divide in one plane only, resulting in a longer filament (*Ulothrix*, p. 45). In branched forms, the cube can occasionally divide in a second plane, giving rise to branches (*Cladophora*, p. 47). More or less equal cell divisions in two planes would result in a blade one cell thick (*Prasiola*, p. 49). Cell division limited to once in one plane, accompanied by equal and many cell divisions in two planes, would result in a blade two cells thick (*Ulva*, p. 50). In some forms the two cell layers become disconnected, resulting in a hollow tube or sac (*Enteromorpha*, p. 52). Finally, the sac may tear, producing a blade one cell thick (*Ulvaria*, p. 51). Such an exercise may not contribute to our understanding of evolution, but it is instructive in appreciating how various forms come about.

Prasiola
Short sea lettuce
Photo 9, p. 26

Prasiola (Greek=green) is a complex of five species that consist of very small blades (at most a few centimetres in length), usually one cell thick. Tufts of these blades resemble little cabbages growing on wood or rock above the high tide level or along freshwater streams. A unique aspect of their distribution is their association with guano of marine birds. Locally, these plants are distributed from Alaska to central California.

Prasiola can tolerate and exploit high-nitrogen conditions that most plants would find toxic. Mike Guiry (University College, Galway, Ireland) described the distribution of this group in Galway as being associated with telephone poles employed by the local canine population, and in one case with a pole behind a public house.

Ulva
Sea lettuce
Photo 10, p. 26

Ulva (Latin=a marsh plant) is represented by eight species, all of which are flat blades, two cells thick. The blade may be long and narrow or fan-shaped, variously lacerated or perforated, and reach a length of 1 m (3'), but is usually less than 30 cm (12"). The colour may vary from very pale to emerald green. Plants exist throughout the intertidal and subtidal regions and in tide pools. This genus is perhaps the most cosmopolitan of all seaweeds, being found on all of the world's coasts. On North America's west coast, *Ulva fenestrata* is distributed from the Bering Sea to Chile.

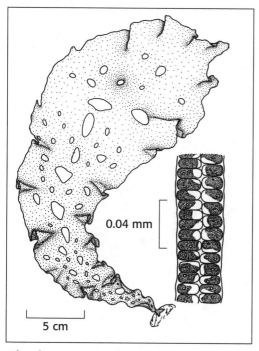

0.04 mm

5 cm

Ulva fenestrata and cross-section (right) showing two cell layers.

Often the margins of these plants are white in contrast to their otherwise rich green. The white portions are cells that have discharged their contents as spores or gametes, leaving only the cellulose cell walls.

Ulva flourishes where there are rich nutrient conditions. Sometimes, growth gets out of hand, resulting in what is called green tide. An extreme example of this phenomenon occurs in Venice Lagoon, Italy, where nutrients from Venice sewage and surrounding farmlands support an extraordinary wild monoculture of *Ulva*. When the waters warm up, the *Ulva* population depletes all available oxygen and begins to rot, resulting in a nauseating rotten-egg smell. Up to 200 metric tonnes (220 tons) of *Ulva* are harvested daily, eight months of the year, to combat this situation. John Merrill (University of Wisconsin) is trying to find other seaweeds of economic value that could compete with *Ulva* for nutrients in Venice Lagoon.

Ulvaria
Dark sea lettuce

Ulvaria obscura represents two bladed forms, each one cell thick: *Kornmannia* (seagrass cellophane), one species, and *Monostroma* (sea cellophane), four species. Blades are fan-shaped and usually less than 30 cm (12") long. These plants are found lower on the beach than most *Ulva* but otherwise are difficult to distinguish in the field. A microscope is required to determine the number of cell layers. Upon drying,

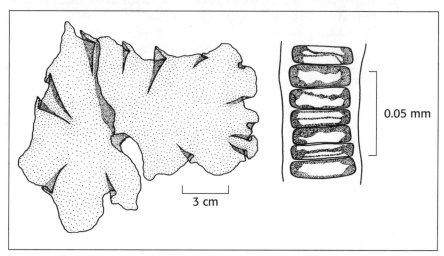

Ulvaria obscura and cross-section (right) showing one cell layer.

Ulvaria turns dark, due to the presence of polyphenol oxidase, an enzyme that is also found in apples and turns them dark upon exposure to air. This feature distinguishes *Ulvaria* from other bladed green seaweeds. *Ulvaria* is distributed from Alaska to northern Washington, and various *Monostroma* species are distributed from Alaska to southern California. *Kornmannia leptoderma* is smaller than *Ulvaria* and *Monostroma*, usually less than 5 cm (2") long, and does not turn dark upon drying. *Kornmannia* grows attached to seagrasses and other seaweeds from Alaska to central California.

Dopamine, an adrenaline-like compound, is another exotic chemical found in *Ulvaria*. Could this sea lettuce give a person an adrenaline rush?

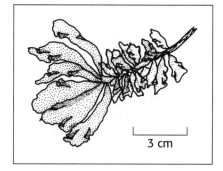

Kornmannia leptoderma on *Zostera*.

Our understanding of the relationships amongst these local bladed forms is largely the result of Maurice Dube's (Western Washington University) efforts. Dube had a green (seaweed) thumb and cultured these plants through their young, informative stages. He died suddenly when accidentally struck by a cyclist. The following poem was read at the Fourth Northwest Algal Symposium, March 1989, in his memory.

Maurice Dube you accidentally died in your sixtieth year,
While walking a darkened street near your home.
In that very special early morning light,
With a tea-quickened pulse and a little stiff,
You and your students greeted many a low tide,
Heralded by the forceful slap of waves.
A multitude of seaweed forms lie exposed,
Accessible to you and your students.
And a glimpse assures the essential presence of your
Rich green undulant *Monostroma*.
Maurice, when the pale, cold, waxing moon
Draws back the watery curtain of tide,
We will join you on those compelling shores,
And once again share the marine wonderment.

—Louis Druehl

Cylinders

Enteromorpha
Green string lettuce

Photo 11, p. 27

Enteromorpha (Greek=intestine form), a complex genus consisting of branched and unbranched tubular plants, is represented locally by seven species. The tubular structure is interrupted as an inflated *Ulva*, where the two cell layers are separated to form a tube. The most common species, *E. intestinalis*, is cosmopolitan, being found on all continents. For one species, *E. linza*, the plant is tubular only at the blade base, otherwise it is identical to *Ulva*. The tubes may reach 50 cm (20") in length but are usually less than 20 cm (8") long, and their width varies from a few millimetres to 4–5 cm (1½–2"). *Enteromorpha*, which is normally distributed throughout the intertidal zone, is very tolerant of fresh water and very salty conditions. Often it grows in freshwater seepages above the high tide level and in high tide pools, which through evaporation become salty to the point of producing crystalline salt. Species of *Enteromorpha* are found from Alaska to Mexico.

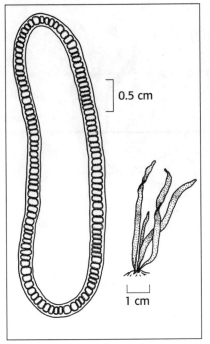

Enteromorpha intestinalis and cross-section (left) showing tubular nature of the plant.

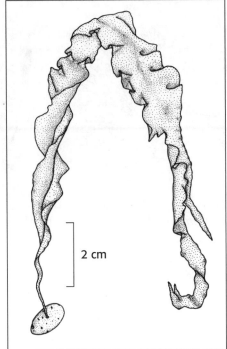

Enteromorpha linza.

Ian Tan and colleagues' comparative DNA studies at the Royal Botanic Garden, Edinburgh, Scotland, have indicated that some species of *Enteromorpha* are more closely related to some *Ulva* species than to other *Enteromorpha* species. This could revolutionize the way we view these two complex genera.

Spongy texture

Codium

Sea staghorn, Green spongy cushion

Photos 12, 13, p. 27

Codium (Greek=animal skin) is a conspicuous genus of five species, all of which are spongy in texture and very dark green. Two basic morphologies exist, branched forms and cushion forms, and both are found in the low intertidal zone from Alaska to Mexico. The branched forms are typified by *C. fragile* (sea staghorn). This species is dichotomously branched (each branch produces two more or less identical branches, forming a "Y" shape), giving rise to plants up to 40 cm (16")

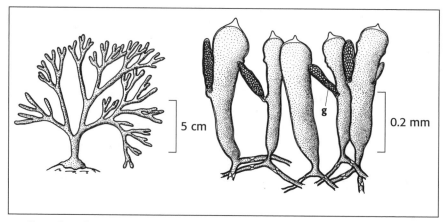

Codium fragile and detail (right) of surface cells with gamete-bearing cells (g) as observed in cross-section.

long. The branches are cylindrical, about 0.5 cm (³/16") in diameter. *Codium setchellii* (green spongy cushion) typifies the cushion form. The low-lying cushion may be up to 2 cm (³/4") thick and 25 cm (10") in diameter.

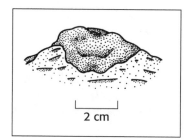

Codium setchellii on a rock.

Codium species consist of long cells (some cells are up to 30 cm/12" long) that branch near the plant surface. There they differentiate into a beautiful scale-like "skin," which can be viewed with the aid of a magnifying glass. These long cells, lacking regular cross-walls, are packed with chloroplast and other cellular components. The sea slug *Alesia* sucks the chloroplasts from *Codium* and houses them in a special gland. Here the chloroplasts photosynthesize and produce a slime with which the animal lubricates its path.

Codium fragile had a unique introduction to the Mediterranean. According to Professor Jean Feldmann (University of Paris), this species was imported from California to the Banyuls-sur-Mer marine station for research purposes prior to World War II. Care was taken to secure these plants in the laboratory, but all for naught. Allied bombers liberated *Codium* and it has persisted as a noxious weed on local swimming beaches ever since.

Unicellular forms

Unicellular green algae are found in two settings on our coasts. Mobile forms, those capable of swimming, are found in tide pools high in the splash zone above the normal high tide level. Other unicellular forms are non-motile and are found inside seaweeds and invertebrates.

Tetraselmis

Tetraselmis (Greek=four branches: refers to presence of four flagella, cellular swimming appendages) consists of six species of swimming unicellular greens that normally inhabit tide pools above the direct influence of the tides. These pools accumulate salt from ocean spray and are subjected to rain and severe evaporation, resulting in severe ranges in temperature and salinity—to the point of being totally dry salt cake! An inhabited pool will resemble pea soup. These forms are found from BC to central California.

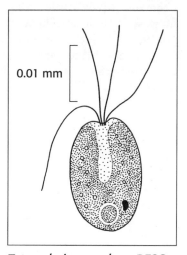

0.01 mm

Tetraselmis maculata. RFSG

Dunaliella

Dunaliella is another swimming unicellular green alga, occupying spray pools along our coasts from northern Washington to central California. *Dunaliella* is distinguished from *Tetraselmis* by having two flagella. It is farmed for its beta carotene, which is sold as an antioxidant for human consumption.

Chlorella-like Plants

Photo 14, p. 27

Chlorella (Greek=green and small) - like unicellular plants (their true identity is not known) are commonly found inhabiting the soft internal tissue of intertidal sea anemones belonging to the genus *Anthopleura*, giving them a distinct green colour. This

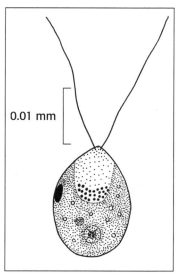

0.01 mm

Dunaliella salina. RFSG

non-motile spherical form is most common in sea anemones in shaded habitats. Gisele Muller-Parker (Western Washington University) has suggested that an important role of this alga, which may provide a portion of the carbohydrates it produces to the nutrition of its host, is to protect the sea anemone from predators. The basis of this protection is not known.

Codiolum, Chlorochytrium

Codiolum (Greek=little ball) and *Chlorochytrium* (Greek=little green fungus) are two unicellular forms generally thought to be the spore-producing phase of several filamentous and bladed green seaweeds. They inhabit bladed and crusty red seaweeds, barnacles and mussels. In the red seaweeds, these minute plants are visible to the naked eye as green dots.

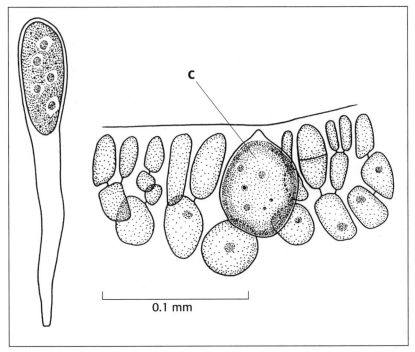

Left: *Codiolum*. Right: *Chlorochytrium*, cross-section of *Schizymenia* showing positioning of *Chlorochytrium* (c). RFSG

Brown Seaweeds

The brown seaweeds are a well-defined group ranging in form and complexity from simple filaments to large, complex kelp. There are no known unicellular brown algae. Technically, the brown seaweeds are not plants. Because of the nature of their flagella and other features, which can be discerned only by electron microscope, they are grouped in a vast kingdom called the Protista (commonly referred to as protists), creatures that are distinct from animals and plants. Other members of the Protista include moulds, many phytoplankton groups, amoebae and parasitic single-celled organisms. All of these are either single-celled or simple filaments and none share the morphological complexity found in the brown seaweeds. The brown seaweeds stand out like an orphaned elephant that has been adopted by a family of mice.

As a group, the brown seaweeds achieve their greatest diversity in colder oceanic waters. Worldwide there are approximately 2,000 species, only a few of which are found in fresh water. Locally, there are approximately 150 brown seaweed species. A conspicuous element of our seaweed flora is the diverse kelp assemblage. In the area covered by this book there are 32 kelp species distributed among 16 genera. This is the richest kelp flora in the world.

Brown seaweeds are brown, brownish, yellowish or blackish, due to their blend of pigments. This blend is dominated by the pigment fucoxanthin, which reflects light in the yellow part of the colour spectrum. Chlorophyll is present, as it is in all photosynthetic organisms, but its relatively low concentration is masked by the fucoxanthin. The opposite occurs in land plants, where chlorophyll is the dominant pigment. The fucoxanthin-like pigments in land plants are only expressed when the chlorophyll is destroyed, giving us autumn colours.

The carbohydrates of brown seaweeds differ from those of the green seaweeds in structural detail that results in a class of compounds that are not digestible by humans. But two unique groups of compounds found in brown seaweeds, algin and fucans, are used in manufacturing consumer goods. Algin is an emulsifier, which has many industrial and food applications, and fucans, the slime of kelp, has potential in the area of medicine (see Utilization and Cultivation of Seaweeds, p. 00).

For the most part, brown seaweeds are easy to distinguish once you become aware of their diversity. Exceptions are those filamentous forms requiring microscopic examination and some closely related species of larger forms.

Crusts

Brown crusts exist either as distinct species or as developmental stages of more elaborate brown algal morphologies. To distinguish these crusts from red crusts, with which they are easily confused, moisten the crust in question. If it appears black or brown it is a brown crust; if it appears red, orangish or purple it is a red crust. In the highest reaches of the intertidal region and extending above the high tide region is a black-crusted lichen that may be confused with brown crusts. However, brown crusts do not occupy the lichen's high position on the beach. Most recently, Megan Dethier (Friday Harbor Laboratories, University of Washington) has advanced our understanding of this important but inconspicuous group.

Ralfsia
Tar spot, Sea fungus

Photo 15, p. 27

Ralfsia (after Ralfs, a British phycologist) consists of several species whose taxonomy is confused. The most common species, *R. pacifica*, is a smoothish, thick crust popularly referred to as the tar spot seaweed. This plant adheres tightly to its rock substrate, even at its margins, and may exceed 20 cm (8") in diameter. It is a prominent crust in the mid-

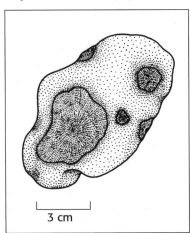

3 cm

Ralfsia pacifica on a rock.

and low intertidal regions from Alaska to Mexico, but does not grow in tide pools, which are often inhabited by another smooth-surfaced form, *R. californica*.

Ralfsia fungiformis is a species distinctive for its overlapping marginal lobes. This lobing gives the plant a fungal appearance. The lobbing pattern and the fact that the crust margins do not adhere tightly to the rock substrate distinguish *R. fungiformis* from *R. pacifica* and *R. californica*. These plants, which may exceed 20 cm (8") in diameter, are found in the mid- and low intertidal regions, while the higher plants are frequently restricted to tide pools. Locally, this species is distributed from Alaska to northern California.

Part of the confusion surrounding *Ralfsia* arises from the discovery that some "species" are really just a developmental stage of other brown seaweeds. For example, *R. californica* has been noted to be the crust phase of a simple brown-bladed seaweed, *Petalonia*. This would lead one to suspect the case is closed—terminate *R. californica*. The distribution of *Petalonia*, however, is much wider than that of *R. californica*. Does this mean a second *Ralfsia* species is implicated in the development of *Petalonia*, or is the crust an environmentally induced phase? See *Petalonia*, small blade forms (p. 65) for further consideration of crust species/erect species complexes.

Analipus
Fir branch (crust phase)

Analipus (Greek=barefoot) is represented by one local species, *A. japonicus*. It produces a persistent crust, which Megan Dethier characterized as consisting of branched, blunt, overlapping lobes. Often this crust supports numerous uprights, which look like fir branches, densely needled. Crusts, often exceeding 20 cm (8") in diameter, are found on bare rock in the mid- and lower intertidal regions, locally distributed from Alaska to Point Conception, California. For further discussion see *Analipus*, the cylindrical brown (p. 72).

Filaments

Brown filamentous forms, like green and red filamentous forms, are often inconspicuous and almost always difficult to identify without the aid of a microscope. Whereas filaments are considered a simple morphology, their range of branching patterns, the structure and positioning of

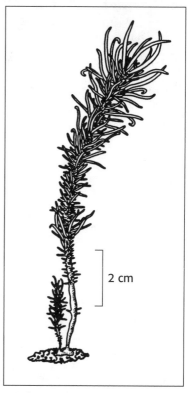

2 cm

Analipus japonicus, showing scaly crust and erect phases.

their reproductive cells and the variety of habitats they occupy signal a sophisticated life form. In our local flora there are about 10 genera of brown filamentous forms, representing about 30 species. These species live attached to rock and other inanimate objects, and to other seaweeds and animals, and some live within the tissues of other brown algae.

Ectocarpus

Ectocarpus (Greek=external fruit) consists of numerous branches reaching 2–3 cm ($3/4$–$1^1/4$") in length and having a fluffy appearance in water. These plants may be found on rock or on other seaweeds (lower intertidal *Fucus* and kelp, pp. 67 and 76, are good places to look). Colonial diatoms may be mistaken for *Ectocarpus* or other filamentous brown seaweeds. Colonial diatoms consist of chains of diatoms, often branching, held together in a common jelly, with the same fluffy appearance and distinctive brown colour as *Ectocarpus*. To distinguish diatoms from brown filaments, rub the plant vigorously between your fingers. If the plant disintegrates, it is a colonial diatom. *Ectocarpus* is found locally from Alaska to Mexico.

Ectocarpus has an isomorphic life cycle with two more or less identical generations. Each generation can produce asexual spores, so that when either generation becomes established in an area, it can reinforce its occupation greatly by multiplying itself through identical genetic copies.

Chemical sexual attractants in brown algae were first discovered by D.G. Müller (Universität Konstanz, Germany), who noticed a pleasant smell emanating from his *Ectocarpus* cultures (a smell, it has been suggested, that was reminiscent of gin). Subsequent study revealed that this perfume was produced by female cells and would attract male gametes. More recently, Müller and his associates have discovered a whole suite of sexual attractants (pheromones) associated with various brown algal groups. For example, one pheromone is produced by representatives of three common kelp families, representing almost 30 genera. The perfume of any kelp female species will attract the sperm of any kelp species. This raises an interesting question: what, if any, are the barriers to inter-species, -genus or -family hybridizations?

3 cm

0.03 mm

Ectocarpus dimorphus and detail (left) of cellular arrangement.

Sphacelaria
Brown rock fuzz

Sphacelaria (Greek=gangrene!?) is represented by six species that are small (up to 6 cm/2^1/2" long, but many only a few mm long). The plants usually occur in small tufts on rock, other plants and sometimes on soft substrate (mud and sand). The gangrene of the name may refer to the pustule-like growths of *Sphacelaria* observed on *Sargassum* in warmer climes. Individual plants within the tufts appear feather-like when splayed out on your finger (a magnifying glass helps). Plants are distributed from Alaska to Mexico in the lower intertidal and subtidal regions.

Microscopically, *Sphacelaria* provides an excellent view of plant organization arising from apical growth. Branches are terminated by distinct apical cells that divide, producing a simple, one-cell wide filament. Cells of this filament continue to divide, giving rise to a regular, simple, multicellular body. Some of the branches modify their development to produce multicellular reproductive spores that look like children's jacks. This form of asexual reproduction involving multicellular spores is rare among the seaweeds.

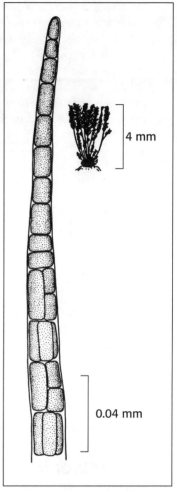

4 mm

0.04 mm

Sphacelaria racemosa and detail (left) showing branch detail with apical cell (at top).

Globular or Cushiony

This morphology is perhaps best represented among the brown seaweeds. It allows the plant to achieve increased surface area, which is important for photosynthesis and nutrient uptake, and at the same time maintain a low profile, thus avoiding some of the stress associated with wave action. These forms are often epiphytic (attached to another organism) but may be found attached directly to solid substrate.

Soranthera
Studded sea balloon

Soranthera (Greek=flowering patches), represented by one species, *S. ulvoidea*, is a dimpled, hollow globe, up to 6 cm (2¹/₂") across, which grows on the bushy red seaweeds *Neorhodomela* (see p. 124) and *Odonthalia* (p. 123). The plants are distinguished by the presence of conspicuous patches (sori) of spore-producing cells, from which this plant gets its name. These patches are much darker than the surrounding tissue. *Soranthera* is distributed in the mid- to low intertidal regions from Alaska to Santa Barbara, California.

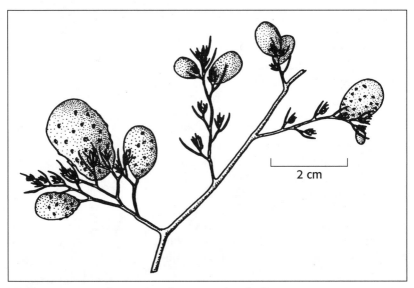

2 cm

Figure 28: *Soranthera ulvoidea* on *Odonthalia*.

Colpomenia
Oyster thief

Photo 16, p. 28

Colpomenia (Greek=to be folded) is represented by two or three local species whose hollow forms vary from a smooth, dimpled sac to clusters of finger-like wrinkled extensions. These light brown plants may reach 10 cm (4") in length. *Colpomenia* is found in the mid- to low intertidal regions from Alaska to southern California, attached to rock and other seaweeds.

Colpomenia's popular name "oyster thief" reflects its ability to lift young oysters to which they are attached and float them out to sea. Many seaweeds, with their ability to retain air, lift their substrate when they reach a critical size. This activity is as damaging, if not more so, for the seaweed as it is for the pirated substrate.

Coilidesme
Sea chip

Coilidesme (Greek=hollowed cluster) is represented by two species whose general form is a fragile, mostly flattened pale brown sac. The plants may consist of many such sacs clustered from a common holdfast. The sacs, which can reach 40 cm (16") in length, may be torn. Plants are attached to rock or the bushy brown seaweed *Cystoseira* (p. 75) in the lower intertidal region from Alaska to Mexico.

Leathesia
Sea cauliflower

Photo 17, p. 28

Leathesia (after Reverend G.R. Leathes, a British naturalist), is represented locally by one species, *L. difformis*. This is an irregular, highly convoluted, low-lying plant whose form suggests the surface of the brain. *Leathesia* is common, conspicuous (up to 15 cm/6" in diameter), and often clustered into yellowish dense patches where it is attached to low-lying seaweeds or rock. Local distribution is in the mid- to low intertidal regions from Alaska to Mexico. When squeezed, *Leathesia* tends to disintegrate, whereas other globular and cushiony forms may pop or wheeze but remain intact.

Small flat blades

These common but mostly inconspicuous forms are distinguished by such anatomical features as number of cell layers observed in cross-section and relative sizes of different cell types. In the field, plant habitat can

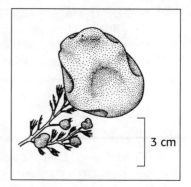

Colpomenia peregrina on *Odonthalia.*

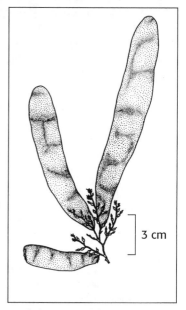

Coilidesme californica on *Cystoseira.*

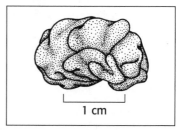

Leathesia difformis.

63

be used to roughly identify these forms. Such identifications should be confirmed by close scrutiny, because we do not know enough about the natural history of these forms (or any seaweeds) to rely on simple environmental associations for positive identification.

Punctaria
Brown sieve

Photo 18, p. 28

Punctaria (Latin=points) is a complex of several species. Representatives are composed of pale brown, elongated blades, usually less than 3 cm (1¹/4") long (rarely up to 20 cm/8"), that arise individually or in clusters from a small disc holdfast. When mature, the blades display scattered little points which consist of hairs and reproductive structures (hence the name *Punctaria*). The common species, which are mostly found growing on seagrasses, are distributed from BC to southern California.

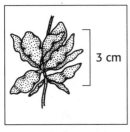

3 cm

Punctaria hesperia on *Phyllospadix*.

Punctaria is locally represented by up to seven species, of which five are distributed from northern Washington to Alaska. This proliferation of species reflects the features used to distinguish them, e.g. size, shape and texture. Invariably, the legitimacy of such distinctions collapses under closer scrutiny. Detailed studies of genetics, life cycles and morphological response to environmental factors may reveal "true" affinities among such species groups. Such taxonomically complex groups arise from the well-intentioned aspirations of taxonomists who see in some unexplained distinguishing feature(s) a new species. Thomas B. Widdowson (gentleman phycologist, University of Victoria) and I suggested changes to the International Code of Botanical Nomenclature, such as self-fumigation of those who, when designating a new species, do not apply to their work the definition of a species that they teach their students. The Widdowson/Druehl study (1973: *Journal of Irreproducible Results*, vol. 20, p. 23) is particularly noteworthy as it was the first international publication from the now world-famous Bamfield Marine Station.

Phaeostrophion
Sand-scoured false kelp

Photo 19, p. 28

Phaeostrophion (Greek=brown twisted cord) is represented by one species, *P. irregulare*, which is distributed from Alaska to Santa Barbara, California. Numerous blades, up to 40 cm (16") long but

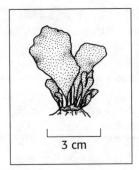

Phaeostrophion irregulare arising from an expansive crust.

usually less than 10 cm (4"), grow from an expansive disc holdfast. The blades are usually dark brown with ragged or irregular ends. Plants are found on rocks that are usually surrounded by sand in the mid- to lower intertidal regions.

Our understanding of the natural history of *Phaeostrophion* comes from the doctoral studies of Arthur C. Mathieson, conducted under the direction of Robert F. Scagel (University of BC, 1966). Mathieson (University of New Hampshire) noted that *Phaeostrophion* is mostly restricted to areas where it remains buried under sand, up to 2 m (6¹/2') deep, from June to November. The branched wiry red seaweed *Ahnfeltiopsis linearis* (p. 121) is represented in this strange habitat also. The sand is transported to an offshore sandbar by winter storms, slowly re-invading the intertidal region through the spring and summer. *Phaeostrophion* can grow away from the influence of sand but competes poorly with other seaweeds. How do these plants survive what might be anoxic (lacking oxygen) conditions for months?

This study and many others have been conducted on the shores of the Point-No-Point Tea House on the west coast of Vancouver Island. This tea house has hosted phycologists from around the world, including the phycological field trip of the Montreal International Botanical Congress (1958). The hostess, Miss Packum, attempted to instruct consecutive generations of Scagelean graduate students in the finer points of British tea. The tea house, now under new management, is still delightful to visit and the shores, once walked on by Captain George Vancouver, remain a phycologist's delight.

Petalonia
False kelp
Photo 20, p. 28

Petalonia (Greek=leaf) is a small (usually less than 10 cm/4" long but may reach 30 cm/12"), light brown blade arising from a small crust-like holdfast. The single recognized species, *P. fascia*, is distributed from Alaska to Mexico, where it is usually found on rock in the mid- to low intertidal regions, but may be found on seagrasses.

Michael Wynne (University of Wisconsin) discovered that photoperiod (the length of day relative to the length of night) and temperature could determine if *Petalonia* would develop into a crust or a blade. Sixteen hours of light and eight hours of dark and warm temperatures (18°C/64°F) resulted in a crust, and eight hours of light and sixteen hours of dark and cool temperatures (10°C/50°F) resulted in a blade. This finding suggests triggers for the seasonal occurrence of crusts (summer) and blades (winter) and demonstrates the plasticity of morphological expression to environmental conditions. Similar responses to long and short daylight conditions have been observed for the blade and crust forms of *Scytosiphon* (see p. 70).

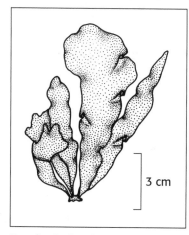

Petalonia fascia arising from a restricted crust.

Michael Wynne's contributions to our knowledge of these small brown algae arose from his doctoral studies at University of California, Berkeley, where he studied with Professor George Papenfuss. His culture studies challenged the accepted belief that static morphology alone could meaningfully define species of the numerous small brown algae. This belief was firmly entrenched. Papenfuss caused a stir when he rejected George J. Hollenberg's report of a filamentous phase in the life cycle of the crust forming brown seaweed, *Hapterophycus*, stating the filaments "probably represented a (contamination)." Such a proclamation from such an eminent phycologist must have been devastating, even if it was incorrect.

I have observed a provocative association between *Petalonia* and *Scytosiphon* on BC beaches. In the mid-intertidal region, *Petalonia* arises from minute crusts/holdfasts; lower, both *Scytosiphon* and *Petalonia* are observed to arise from what appears to be the same crust/holdfast; and lower still, only *Scytosiphon* is observed. Is this a coincidence or a signal of great discoveries to be made?

Dichotomously branched

When a seaweed branches dichotomously it produces two more or less equal branches, directy opposite each other, looking much like the letter "Y." As this process is repeated, the plant assumes an overall flattened appearance. The branches may be flattened or rounded in cross-section. This form is positioned in the sea very differently from the radially branched seaweeds. In the dichotomous system there is a short distance from the outer boundary of the plant to the innermost branches, resulting in less self-competition for light and nutrients.

The common dichotomously branched species that inhabit the Pacific coast belong to the order Fucales—robust plants usually occupying the mid- to high intertidal region—or the order Dictyotales, thin plants usually occupying the low intertidal and subtidal regions.

Fucus
Rock weed

Photo 21, p. 29

Fucus (Greek=seaweed) is the most common and commonly known seaweed in cold northern hemispheric waters. Plants reach 40 cm (16") in length, are profusely branched and have a conspicuous midrib running the length of the flattened branches. On older branches the tissue surrounding the midrib may be eroded, leaving only the midrib. The branch tips are often inflated, buoying the plant upright in the water. When mature, these tips contain the conceptacles (cavities) that bear the egg and sperm. The plants are found from the mid- to high intertidal regions from Alaska to Point Conception, California. They are most successful in wave-sheltered to moderately wave-exposed situations. In fully wave-exposed situations they are replaced by other, similar forms.

Two species are recognized locally. *Fucus gardneri* is most common throughout this area of the Pacific. *Fucus spiralis* has only recently been recognized from Alaska to northern Washington. *Fucus spiralis* has a flange ringing the inflated branch tip and a weakly developed system of midribs, whereas *Fucus gardneri* lacks the ring flange and has a well-developed system of midribs. When the two species co-inhabit the beach *F. spiralis* is the higher form.

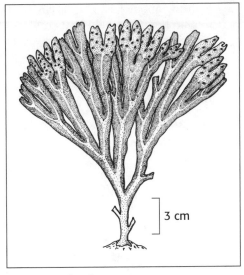

Fucus gardneri.

3 cm

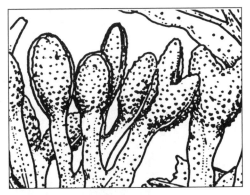

Fucus spiralis, detail of branch tip showing distinctive flange. WRT

The life cycle of *Fucus* and other fucoids is unique in the plant world in that it closely follows that of animals, where the "vegetative" body directly produces eggs and sperm. Most plants have an intervening generation that produces eggs and sperm or their plant equivalents. Our understanding of sexuality in seaweeds probably started with J. Stackhouse, a British naturalist. In the late eighteenth century he observed fertilization in *Fucus* and noted it was accompanied by a subtle vapour (see *Ectocarpus*, p. 60). He went on to criticize earlier researchers who held that seaweeds were without sex, a view he held as "the most unphilosophic that I could have expected to have met in this enlightened age." I wonder how kindly future seaweed researchers will treat us and our dogma.

Pelvetiopsis
Dwarf rock weed
Photo 22, p. 29

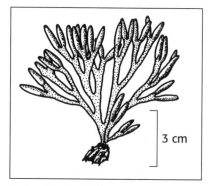

Pelvetiopsis limitata.

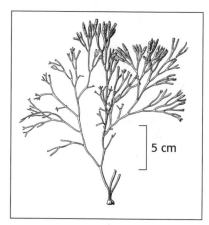

Silvetia compressia. AH

Pelvetiopsis (after Dr. Pelvet, a French naturalist) is represented by two species. One, *P. limitata*, is common on wave-exposed shores from BC to central California. *Pelvetiopsis* closely resembles *Fucus* but has no midrib, is usually smaller (less than 10 cm/4" long) and is oval in cross-section. It occupies the same intertidal position as *Fucus* but does better than *Fucus* in fully wave-exposed areas. A similar fucoid, *Silvetia* (Silva's rock weed), extends farther south, from central California to Mexico with an isolated population in BC. *Silvetia* (in honour of Paul Silva, University of California, Berkeley) is larger (usually longer than 30 cm/12") than *Pelvetiopsis* (usually shorter than 15 cm/6") and differs from *Pelvetiopsis* in that it produces eggs in pairs rather than solo. *Silvetia* was previously named *Pelvetia*.

I recently discovered (or rediscovered) *Silvetia* on the southeast shores of Vancouver Island in front of the Pacific Biological Station. This species had been reported earlier from BC but the claim had been discredited as there were no supporting specimens. The disjunctive, or broken, distribution of *Silvetia* from central California to BC may be the result of an accidental introduction. Another possible explanation is that the BC population is a relict from the last glacial period, having somehow avoided the ice covering. The last glacial period is thought to have reached only as far south as northern Washington.

Hesperophycus

Hesperophycus (Greek=evening star seaweed) consists of one species, *H. harveyanus*, which closely resembles *Fucus* in its branching pattern, in having a midrib, in having the same size range (up to 50 cm/20" long) and in its intertidal positioning. It may be distinguished from *Fucus* by parallel rows of pits running along both sides of the midrib. *Hesperophycus* is distributed from Monterey to Mexico.

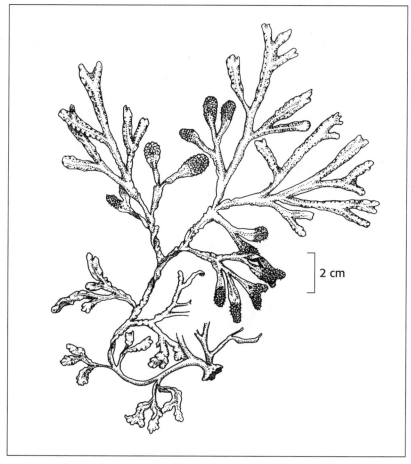

2 cm

Hesperophycus harveyanus. Note pits along some branches. AH

Pachydictyon

Pachydictyon (Greek=a thick net)/*Dictyota* (mermaid's gloves) (Greek=net-like) are each represented locally by one species, *P. coriaceum* and *D. binghamiae*. These forms are members of the order Dictyotales. They are thin, fairly uniformly dichotomously branched plants that may reach 40 cm (16") in length. *Pachydictyon* is thicker and darker than the more delicate *Dictyota*. *Dictyota* often has marginal spines; *Pachydictyon* does not. *Dictyota* is found from BC to Mexico and *Pachydictyon* from Oregon to Mexico. Both forms are found in the lower intertidal and subtidal regions.

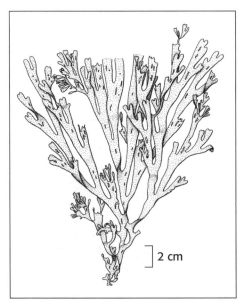
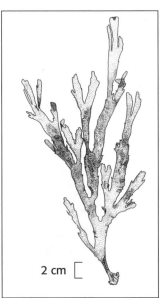

Left: *Pachydictyon coriaceum*. Right: *Dictyota binghamiae*. AH

Cylinders

The cylindrical morphology is expressed in branched and unbranched brown seaweeds. Many brown seaweeds have cylindrical parts, but only those that are cylindrical throughout are included here.

Scytosiphon
Soda straws

Photo 23, p. 29

Scytosiphon (Greek=whip-like tube) represents an unbranched, hollow tubular morphology. Plants are less than 1 cm (^{1}/2") in diameter, may reach 50 cm (20") in length and are usually clustered. Two species are recognized in our area. The most common species, *S. simplicissimus*, is distinctive in that the tube is pinched like sausage links over its length. Plants are typically pale brown. Local distribution is from Alaska to Mexico, where plants are mostly restricted to the lower intertidal region. (See *Petalonia*, p. 65, for *Scytosiphon* life-cycle considerations.)

Melanosiphon
Twisted soda straws

Melanosiphon (Greek=dark tube), a reddish-brown, twisted tube reminiscent of a unicorn's horn, is represented by one species, *M. intestinalis*.

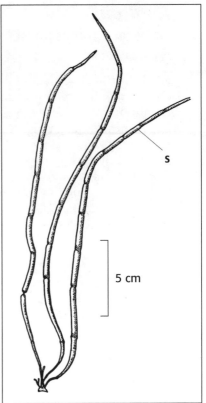

Scytosiphon simplicissimus. Note pinching of the tubes.

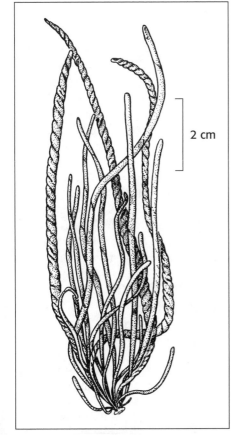

Plants are often clustered and may reach 15 cm (6") in length. This species is restricted to the mid- to high intertidal regions, often associated with the lips of high intertidal tide pools, from Alaska to central California. Until recently this species was known as *Myelophycus intestinalis*.

Melanosiphon intestinalis. AH

Analipus
Fir branch (erect phase)

Photo 24, p. 29

Analipus (Greek=barefoot), the erect phase arising from a conspicuous crust, consists of a main branch surrounded by numerous uniformly short branches (1–5 cm/$^1/_2$–2" long), which are arranged much like fir needles on a twig. Numerous branches up to 30 cm (12") long may arise from a scaly crust (see Crusts, p. 59). The erect phase persists for only part of the year (ephemeral) but the crust may persist for several years. *Analipus japonicus*, the only recognized species, is found in the mid- to high intertidal region from Alaska to Point Conception, California.

The Japanese know this seaweed species as matsumo, an ingredient that may be incorporated in a great number of dishes (see Nutrition and Cooking, p. 153). It is one of the few seaweeds that is eaten fresh and is frequently used in salads.

Saundersella

Saundersella (De A. Saunders, an American phycologist) is an infrequent visitor of *Analipus*. The single species *S. simplex* is an unbranched cylindrical epiphyte that mimics the small branches of *Analipus*, but can be distinguished from them by its greater length (5–10 cm/2–4"). *Saundersella* is known from Alaska to northern Washington. Discovery of this plant is considered a major achievement for emerging phycologists!

Desmarestia
String acid hair
(cylindrical form)

Photo 25, p. 29

Desmarestia (after A.G. Desmarest, a French phycologist) consists of large conspicuous plants that may be cylindrical throughout or markedly

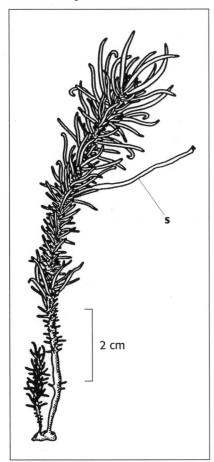

s

2 cm

Analipus japonicus with *Saundersella simplex* (s).

flattened. Cylindrical forms are profusely branched, often wiry in appearance, up to 3 mm (1/8") in diameter, and may reach 1 m (40') in length. Plants are attached to their normal rock substrate by a minute disc-like holdfast. Three species of cylindrical forms are recognized. *Desmarestia aculeata* has mostly alternate branching (lateral branches staggered along the main axes), and *D. kurilensis* and *D. viridis* have opposite branching (lateral branches paired along the main axes). If the plant is cylindrical throughout, it is *D. viridis*, if not it is *D. kurlenensis*. *Desmarestia kurilensis* represents a transition between the cylindrical and flattened forms. *Desmarestia* occupies the low intertidal and subtidal regions from Alaska to Mexico.

Desmarestia aculeata.

Radially branched with some parts flattened

Radial branching is characterized by having side branches arising all around the main axes, usually in some regular fashion. This is a common form of branching for many land plants and is most conspicuous in various conifers. The radially branched plant maximizes its occupation of its three-dimensional space but may suffer from self-shading and having only the outer branches in a nutrient-rich habitat.

The radially branched species grow by means of apical cells (responsible for apical growth), that terminate each branch. The actively dividing apical cells lay down the body of the branch and give rise to lateral branches. The arrangement of the lateral branches is referred to as phyllotaxy. For example, the plants described below have a 2/5 phyllotaxy: every fifth branch is positioned directly below the fifth branch above and it takes two spirals to achieve this positioning. This application of phyllotaxy to the plants described below represents the essence of my Master's thesis (University of Washington, 1961). When you find yourself bored—say, waiting for a bus—eyeball down a branched plant and determine its phyllotaxy. When you get good at this pastime, try determining the phyllotaxy of scales on a pine cone or a pineapple.

Sargassum
Japanese weed
Photo 26, p. 30

Sargassum (Spanish=seaweed, particularly free floating) is a very complex genus consisting of more than 150 species, most of which are restricted to warm waters. *Sargassum* is characterized by having lateral branches arising from the axils of basal blades (the elbows of the blades and the stems from which they arise) and having smooth, rounded floats. The plants are very profusely branched, reaching up to 10 m (33') in length, but usually 1–3 m (3'–10'). When the plant is fertile,

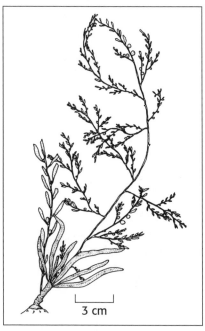

3 cm

Sargassum muticum.
Note rounded floats.

cigar-shaped appendages (receptacles) are evident among the floats. These structures bear conceptacles (pits) that contain the eggs and sperm. The base of the plant, which is perennial, is darker and thicker than the more ephemeral terminal branches. This base often supports "buds," which expand and develop into new terminal systems. These plants are found from Alaska to Mexico, usually in the upper subtidal region, but also scattered through the lower intertidal region and in tide pools. Frequently they form very dense beds. They are most common in wave-sheltered areas, but they can be found everywhere.

The local species, *S. muticum*, was introduced from Japan along with oyster spat, probably as early as 1902. Due to its similarity to *Cystoseira* (see below), however, it wasn't recognized as a distinct entity until 1951 when Dorothy Fensholt described it in her doctoral dissertation at Northwestern University. The species has adapted well to local waters, occupying a much wider range of situations along this coast than on its home turf in Japan. In 1972 I read that oyster spat was to be transplanted from BC to Europe and predicted in a letter to the journal *Science* an invasion of *S. muticum* in the Atlantic. Two years later, an article in *Nature* proclaimed me correct. So there! Its spread through western European waters has been very rapid, and in Great Britain the citizenry have been called upon to eradicate this invader—without success.

Cystoseira
Bladder leaf

Cystoseira (Greek=a chain of floats) is represented by two local species, *C. geminata*, which is distributed from Alaska to Oregon, and *C. osmundacea*, distributed from Oregon to Mexico. Both species are profusely branched and may reach lengths of 8 m (26') but usually are much shorter. They are distinguished from *Sargassum* by having lateral branches arising directly from the main axes without a basal blade and having pointed floats. *Cystoseira osmundacea* has chains of floats and *C. geminata* has solitary floats. Plants are found in the lower intertidal and upper subtidal regions, often in moderately wave-exposed areas.

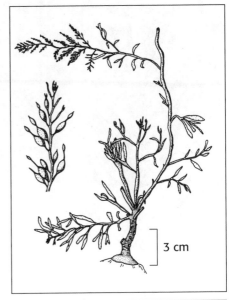

3 cm

Feather-like branching

Desmarestia
Flattened acid leaf

Photo 27, p. 30

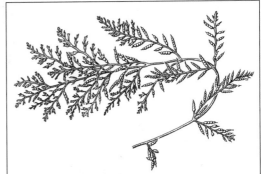

Desmarestia species are either flattened or cylindrical (see p. 72). The flattened forms have conspicuous central axes with regular rows of lateral branches,

Top: *Cystoseira geminata* and detail (left) showing pointed floats. Bottom: *Cystoseira osmundacea*, showing chains of pointed floats. AH

whose arrangement is reminiscent of hairs on a feather. This flattened condition is referred to as being ligulate, hence *Desmarestia ligulata*. A faint branching midrib system may be evident, running the length of the branches. Overall, the plants may exceed 1 m (3') in length. Plants are attached to rock by a disc-like holdfast in the lower intertidal and subtidal regions from Alaska to Mexico.

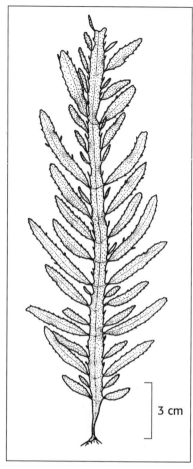

Desmarestia ligulata.

Up to ten species/forms of flattened *Desmarestia* have been recognized for our region. A. R. O. (Tony) Chapman (Dalhousie University, Halifax) investigated morphological relationships of these during postdoctoral work with Robert Scagel. From this study, he concluded there was strong morphological overlap of distinguishing features among the various species and he proposed to reduce their number to three. However, several researchers elsewhere have noted that narrow (less than 1 cm/^{1}/2" wide) species are monoecious (male and female structures on the same plant), whereas wider forms are dioecious (male and female structures on separate plants). Currently, six species are suggested to exist in this area.

Professor B.J.D. Meeuse (University of Washington) noted *Desmarestia* produces and stores free sulfuric acid, giving the plants a pH of 0.8–1.8, which is similar to human gastric juices. When plants are handled, they often release this acid, "digesting" themselves and any life in contact with them. If you must collect these plants, keep them separate from other organisms. Dr. Meeuse had a unique research orientation. He was equally at ease studying the biology of pollination, the biochemistry of heat-producing plants and any esoteric biochemical phenomenon that presented itself. To the chagrin of generations of botany graduate students, be they ecologists, taxonomists or phycologists, he insisted they have a basic comprehension of biochemical processes. The range of his interests and the depth of his understanding of biological phenomena are enviable.

Kelp

Kelp are the major floral component of our intertidal and subtidal shores. They form the architecture and food base for the associated plant and animal communities. Taxonomically, they are all closely related (all belong to the order Laminariales) and local forms share similar patterns of growth, morphological differentiation and life cycle.

Growth is accomplished by means of an intercalary meristem, an area of active cell division and tissue formation located between the base of a blade and the top of a stipe. This meristem may contribute more to the stipe relative to the blade, producing a stipe-dominated plant, or it may produce more blade tissue. The meristem may regularly split, resulting in a branched plant, or it may not split, producing a single blade. Basically, all kelp consist of a holdfast or attachment organ, a stipe or stem-like structure, and a blade or leaf-like appendage. The holdfast may be disc-shaped or composed of numerous haptera (finger-like projections). The stipe may be branched or not, solid or hollow, rigid or flexible and long or short. The blade may be split into finger-like segments or be unsplit, having ribs or folds or being smooth with or without small bullations (undulations). In some species spore production is restricted to special blades called sporophylls, which are arranged along the stipe, and in other species spore production takes place on vegetative blades. Using all of the above variables, hundreds of distinctive morphologies are possible. However, in reality we have only about 32 distinctive local species and perhaps 100 species worldwide.

The life cycle of kelp alternates between the large spore-bearing plants we regularly see and microscopic plants that produce eggs or sperm. These two different morphologies experience very different environmental conditions while existing in the same locality. The large spore-producing plants must withstand the stress of water movement and, because they are fleshy, they have greater demands for solar energy and nutrients. The more cryptic habitat allowed the microscopic sexual plants shelters them from the harshest water motion but at the same time puts them in close competition with neighbours for light and perhaps nutrients. Their filamentous forms give them the advantage that all cells are photosynthetic and capable of taking up nutrients, whereas in the fleshy spore-producing plants, only the surface cells are capable of taking up nutrients and photosynthesizing.

Simple blades

The expression "simple blades" is used to set apart those kelp lacking midribs, folds, sporophylls and branching. In reality, these plants may have blades that are variously sculptured by ridges and valleys or dissected into finger-like segments. These details in blade morphology may strengthen the blade or cause turbulence in the water surrounding the blade. This latter function is a means of replacing stagnant nutrient-depleted water adjacent to the blade with nutrient-rich sea water. This problem may arise when the plant is subjected to very little water motion.

Laminaria
Tangle

Photos 28, 29, p. 30

Laminaria (Latin=thin leaf) is represented worldwide with about 30 species. Locally, nine species are recognized. These species are placed in subgroups on the basis of holdfast and blade morphology. The following dichotomous key establishes four groups to accommodate local species.

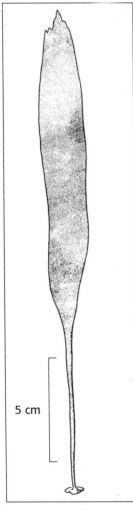

5 cm

Laminaria ephemera, showing disc holdfast.
AH

1. Holdfast disc-like: *L. ephemera, L. yezoensis.*
   Holdfast composed of branched, finger-like projections: Go to 2.

2. More than one stipe/blade complex arising from holdfast: *L. sinclarii, L. longipes.*
   Only one stipe/blade complex arising from holdfast: Go to 3.

3. Blade dissected into numerous, regular segments: *L. setchellii, L. dentigera.*
   Blade not dissected into numerous, regular segments: Go to 4.

4. Blade with longitudinally oriented corrugations: *L. farlowii.*
   Blade smooth or with lateral rows of corrugations: *L. saccharina, L. groenlandica.*

Laminaria ephemera and *L. yezoensis* are two very different species sharing a disc-like holdfast. *Laminaria ephemera* is a delicate, short-lived (ephemeral) plant, up to 2 m (6½') long with a blade usually less than 10 cm (4") wide. Plants are found from Alaska to central California in the lower intertidal and upper subtidal regions. *Laminaria yezoensis* is a tough, perennial species with a wide, often torn blade that is usually wider than 20 cm (8"). This species is found from Alaska to Vancouver Island in the low intertidal region. Often it is found among *L. groenlandica.*

    Laminaria sinclarii and *L. longipes* are two very similar perennial species that are distinguished by anatomical detail. They consist of numerous rigid stipes terminated by narrow (usually less than 5 cm/2" wide) blades of varying length. *Laminaria sinclarii* is found from central BC to central

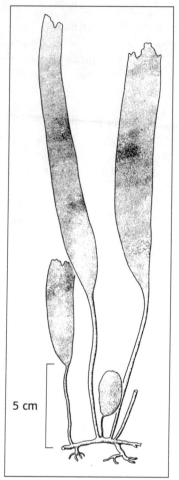

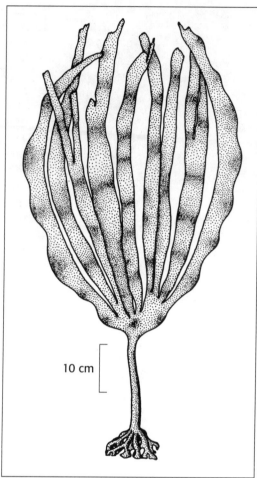

Laminaria sinclairii, showing rhizome-like holdfast. AH

Laminaria setchellii, showing finger-like holdfast and regular splitting of the blade.

California, where it is found in the low intertidal region, often associated with sand. *Laminaria longipes* is known from lower intertidal stands in Alaska and from a subtidal population near San Juan Island, Washington.

Laminaria setchellii and *L. dentigera* are two similar species, each having a rigid stipe (up to 50 cm/20" long) and broad (usually greater than 30 cm/12" wide) blades that are digitate (dissected into numerous fairly uniform segments). They may be distinguished by the depth of splitting on the blade. *Laminaria setchellii* is deeply split (almost to the blade base), whereas splits in *L. dentigera* stop at least 10 cm (4") above the blade base. Both species occur in the lower intertidal and upper subtidal regions, usually in wave-exposed areas, with *L. dentigera*

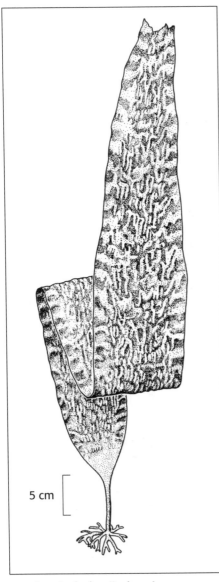

Laminaria farlowii, showing longitudinal grooves. AH

5 cm

known in Alaska only and *L. setchellii* being distributed from southern Alaska to Mexico. Much of the literature erroneously labels digitate *Laminaria* from California as *L. dentigera*.

L. farlowii may reach 5 m (16^1/$_2$') in length (usually less than 1 m/3'). The blade is unique in being sculptured by numerous, more or less regular and abrupt, longitudinal undulations. Plants are found in the lower intertidal and upper subtidal regions from central California to Mexico with a suspected population near Vancouver Island.

Laminaria saccharina and *L. groenlandica* are similar species whose blades may be entire or dissected into a few irregular segments that are the result of tearing. The blades may bear regular undulations or be smooth. Both species are distributed from Alaska to central California in the lower intertidal and/or subtidal regions. The two species are positively distinguished only by anatomical details. They may be distinguished in the field by habitat: the intertidal distribution of *L. saccharina* is restricted wave-sheltered areas and *L. groenlandica* is restricted to moderately to fully wave-exposed areas. However, I have seen both species growing together on the Friday Harbor Laboratories floats (San Juan Island, Washington).

Laminaria groenlandica is conspicuously darker and thicker than *L. saccharina*. Rae Hopkins (Canadian Kelp Resources Ltd., Bamfield) describes *L. groenlandica* as looking like dark Belgian chocolate and *L. saccharina* as milk chocolate. The taxonomic status of *L. groenlandica* is considered confused and some have named this species *L. borgardiana*.

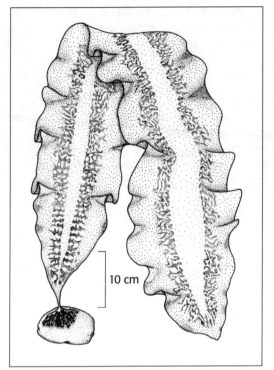 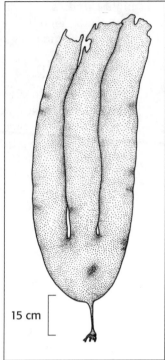

Figure 53 & 54: Left: *Laminaria saccharina*, showing a wrinkling of the blade. Right: *Laminaria groenlandica*, showing irregular tearing of the blade.

Hedophyllum
Sea cabbage

Photo 30, p. 30

Hedophyllum (Greek=seat-shaped leaf) has only one species, *H. sessile*. The plant is essentially stipe-less, with a single blade arising more or less directly from a highly branched holdfast. The blades vary from very lacerated, smooth forms restricted to less than 20 cm (8") long in very wave-exposed areas to large (up to 80 cm/32" long), little-dissected and

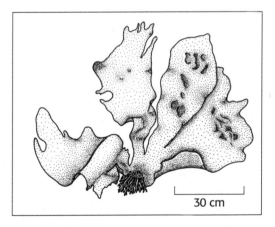

Hedophyllum sessile. Note absence of a stipe.

81

highly undulate forms in wave-sheltered areas. The sheltered morphology is reminiscent of elephant ears. *Hedophyllum* is restricted to the mid-intertidal region from Alaska to Monterey.

Winged kelp

The winged kelp are characterized by having sporophylls (special reproductive blades) in a pinnate arrangement (along the sides of the stipe in much the same pattern as hairs on a feather). *Lessoniopsis* and *Macrocystis* are kelp genera that also bear sporophylls but their sporophylls are not conspicuous. These two genera are treated below (see Repeatedly branched kelp, p. 89).

Kelp demonstrate various stages of reproductive specialization. All species produce spores in concentrated patches called sori (singular: sorus) that are composed of spore-producing cells and cells shaped like umbrellas, which protect the spore-producing cells. Studies have strongly suggested that these patches are further protected with high concentrations of phenolic compounds that act as anti-grazing agents—a form of chemical warfare. Some kelp have sporophylls bearing the sori, leading to a division of labour of the plant parts. An advantage of this strategy is that the sori are independent of the actively growing vegetative blades, which are constantly being renewed. This allows a sorus to persist longer than it would if it were restricted to the vegetative blades. In some instances, plants may produce sori on vegetative blades and vegetative tissue. This is common in *Macrocystis* and in *Alaria tenuifolia*. Cathy Pfister (University of Chicago), while studying with Robert Paine (University of Washington), induced sori on the vegetative blades of *Alaria nana* by experimentally removing the plant's sporophylls.

The same progression of adapting leaves to accommodate reproductive processes occurred in flowering plants. Here, the leaves concentrated reproduction in certain parts of the plant (cones, flowers), protected the reproductive cells by wrapping around them (stamens, pistils), and made them attractive (petals). These processes led the famous German poet Johann Wolfgang von Goethe (*Metamorphose der Pflanzen*, 1790) to proclaim "*Alles ist Blatt*" (Everything is leaves) when describing the flower.

Alaria
Winged kelp

Photos 31, 32, p. 31

Alaria (Latin=wing) is represented locally by seven quite similar species. Each consists of a branched holdfast that gives rise to a stipe up to 60 cm (24") long but usually less than 30 cm (12"). Arranged

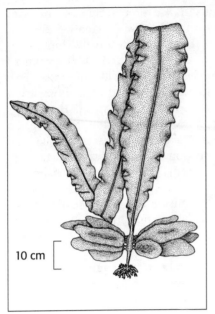
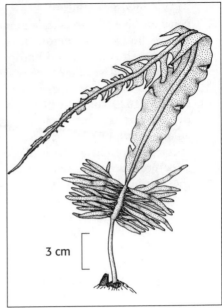

Left: *Alaria marginata*. Right: *Alaria nana*.

along the upper third of the stipe are two rows of sporophylls of various sizes and shapes. The blade varies in length from 30 cm (12") to several metres (most are less than 2 m/6$^{1}/_{2}$') and has a characteristic midrib running its length. Species of *Alaria* have been defined mostly on the basis of their sporophyll and stipe dimensions by Thomas B. Widdowson in his PhD dissertation (University of BC, 1964). The more common species may be distinguished using the following dichotomous key derived from his study.

1. Midrib hollow; blades, which are buoyed to the surface, often longer than 10 m (33'). Found subtidally, Alaska and northern BC: *A. fistulosa*.
   Midrib solid, plants usually less than 3 m (10') long: Go to 2.

2. Stipe longer than 15 cm (6"), sporophylls of irregular shape. Found in lower intertidal and subtidal regions from Alaska to Washington: *A. tenuifolia*.
   Stipe shorter than 15 cm (6"), sporophylls of regular shape: Go to 3.

3. Sporophyll length less than 5 times its width. Found in the lower intertidal and subtidal regions from Alaska to Monterey, California: *A. marginata*.
   Sporophyll length more than 5 times its width. Found in the mid-intertidal region from Alaska to Oregon and possibly to Monterey: *A. nana*.

Widdowson's study of the genus *Alaria* represented the birth of computer-based, large-scale numerical taxonomy. During his doctoral studies he had the onerous task of travelling extensively throughout Asia, North America and Europe, where he collected 5,000 specimens of *Alaria* and studied type specimens in the world's herbaria. Two thousand of his specimens were subjected to statistical analyses using an IBM 1620 computer (punch cards and lots of time). His results demonstrated considerable morphological variation and overlap among previously described species. The means of describing the local species represent a reasonable dissection of the statistical data (e.g., stipe greater or less than 15 cm/6" long). The actual relationships of these dubious species may be resolved through DNA studies.

Alaria fistulosa deserves special consideration due to its hollow midrib. The blades of this species, which may reach 30 m (99') in length and 1 m (3') in width, are buoyed by this midrib to the surface, producing a substantial canopy. This is the only canopy-producing kelp in the northwest Pacific, where it is restricted to the waters north of Japan. On our shores, *A. fistulosa* is found throughout Alaska. It has been reported in northern BC, but its persistence there is questionable.

Pterygophora
Walking kelp

Photo 33, p. 31

Pterygophora (Greek=bearing wings) is represented by one species, *P. californica*, distributed from Alaska to Mexico. The plant consists of a

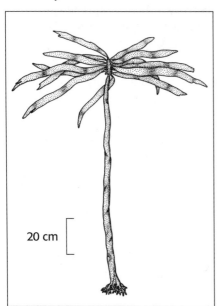

woody, stiff stipe (up to 2 m/6½' in length), terminated by a relatively small vegetative blade with an inconspicuous midrib. The upper reaches of the stipe support two rows of spore-bearing blades along its margin. This species is subtidal; however, shallow plants often project above the water surface during low tide.

In the Northwest, aboriginal people knew *Pterygophora* as the walking kelp, a reference to the movement of these plants through the shallow subtidal region. *Pterygophora* often becomes established on cobble in wave-exposed areas. As the plant

20 cm

Pterygophora californica.

grows, it obtains a critical size where the sweep of the waves is sufficient to bounce the plant and its attached cobble about. Natives would arrange these plants, with their attached cobble, in rows to deflect salmon into fish traps.

Plants may live as long as twenty-five years. This was determined by Rob DeWreede and Terrie Klinger (University of BC) by first establishing the validity of "growth rings" produced by the stipe and then sampling subtidal populations.

Egregia
Feather boa

Photos 34, 35, p. 31

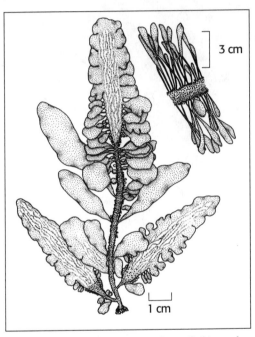

Egregia (Latin=remarkable), represented by one species, *E. menziesii*, is a large (up to 15 m/49^{1}/$_{2}$' long), profusely branched and confusing plant. Its tight, gnarly holdfast anchors numerous branches that seem to arise from an initial branch as randomly positioned outgrowths. The narrow branches support innumerable lateral appendages along their margins. Some of these are floats; others are variously shaped small blades, some of which (sporophylls) bear spore patches (sori). Plants are distributed from Alaska to Mexico in the low intertidal and subtidal regions.

Egregia menziesii, young plant (left) and detail of old plant.

Judging from its geographic distribution, abundance and growth performance, *Egregia* seems to represent relatively warm water kelp. Over the past few years I have noticed a considerable increase in its population and density along the west coast of Vancouver Island. This enhancement has taken place in areas normally dominated by the kelp *Alaria* and *Hedophyllum*, two cold-water genera. Could this invasion signal a general warming of our waters, creating conditions more favourable for warm-water forms over cold-water forms?

Egregia forms dense canopies in local waters. These forests are usually in shallower water than the kelp canopies created by *Nereocystis* and *Macrocystis*.

Eisenia
Forked kelp

Photo 36, p. 32

Eisenia (after G. Eisen, a German scientist) is represented by one local species, *E. arborea*. This species consists of a stout stipe (up to 1.5 m/5' long), terminated by two tufts of sporophylls (spore-producing blades). The sporophylls are arranged in scrolls, the youngest ones

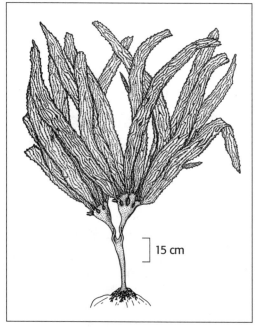

being mere spikes and becoming progressively larger and more distinct. The mature sporophylls (up to 40 cm/16" in length) are frequently ruffled and have distinct teeth along their margins. Populations of this species are found in BC and then, more or less continuously, from Monterey, California, to Mexico. The species is subtidal but projects above the sea surface at low tide in some localities.

15 cm

The disjunct (disrupted) distribution of *E. arborea* along our shores raises some interesting questions. How might such a distribution arise? Do the northern plants represent a relict population that survived during the last

Eisenia arborea.

glaciation in coastal refugia? Or do they represent plants introduced with beach ballast from southern California during the frantic sea otter fur trade era (1740s–1820s)?

Eisenia is represented by a few poorly understood species. On both sides of the Pacific, *Eisenia* is the southernmost kelp representative. A species of marine iguana inhabiting the Galapagos Islands feeds on *Eisenia*.

Ribbed kelp

Ribbing is a common feature among kelp species. Close examination reveals that what are referred to as ribs are really two distinct morphologies. In true ribs, the ribbed portion of the blade is thicker than the adjacent tissue. In other instances, what appears to be a rib is in reality a fold, which creates a raised portion on one side of the blade and a depressed portion on the opposite side, rather than a significantly thicker blade portion.

Ribs are more pronounced in wave-exposed areas than in sheltered areas, which suggests that ribs and folds probably function to strengthen the blade much as the structural ribs strengthen a boat. Further, the ribs may stabilize the plant as it experiences severe water motion, much as the keel stabilizes a boat.

Pleurophycus
Broad-rib kelp
Photo 37, p. 32

Pleurophycus (Greek=ribbed seaweed) consists of one species, *P. gardneri*, which is distributed from Alaska to northern California. Plants are up to 1.5 m (5') long, consisting of a branched holdfast, a stout stipe (up to 50 cm/20" long) and a blade supporting a pronounced single midrib running its length. When the plant is fertile, dark, conspicuous sori (spore patches) are evident on the midrib.

Pleurophycus occurs in the lower intertidal and shallow subtidal regions in areas subjected to considerable water motion. Frequently it is encountered scattered among *Laminaria setchellii*. First-year plants can be easily confused with *Laminaria* (see p. 78), as they do not have midribs.

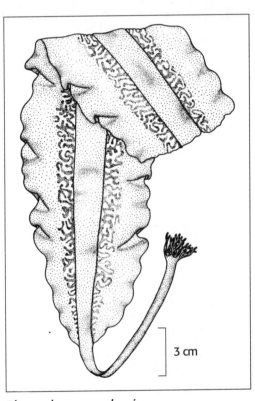

Pleurophycus gardneri.

Cymathere
Triple-rib kelp

Cymathere (Greek=wavy) is represented locally by one species, *C. triplicata*, which is distributed in the low intertidal and subtidal regions from Alaska to northern Washington. *Cymathere* is an annual plant

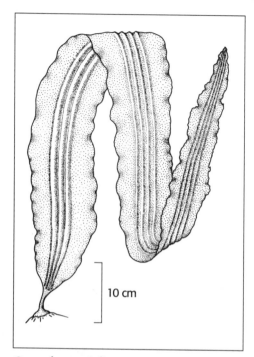

Cymathere triplicata.

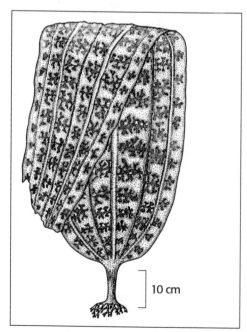

Costaria costata.

with a disc-like holdfast, a short cylindrical stipe and a blade (up to 1.5 m/5' long) with three longitudinal folds.

Costaria
Five-rib kelp

Photo 38, p. 32

Costaria (Latin=rib) consists of one species, *C. costata*. The plant's branched holdfast gives rise to a stipe (up to 50 cm/20" long but usually less than 30 cm/12"). The blade, up to 2 m (6 1/2') long, has five midribs running its length, three on one side and two on the other. The midribs are convex on the top and concave on the bottom, similar to folds, but they are thicker than the surrounding tissue. The blade tissue between the ribs is elaborately contorted into a system of ridges and valleys. *Costaria* is an annual, distributed from Alaska to southern California in the low intertidal and upper subtidal regions.

Costaria displays a remarkable range of shapes, reflecting the degree of wave exposure it encounters. In wave-exposed sites the plants are narrow and thick. Their stipes are ridged and a series of regular perforations run the length of the blade. Wave-sheltered plants are broad and thin, having smooth stipes and no perforations. Transplant studies have shown these differences in morphology to be environmentally induced. For example, a plant moved from a wave-exposed locality to a sheltered area will produce new blade tissue characteristic of wave-sheltered plants. This morphological response to environmental differences is called phenotypic plasticity.

Repeatedly branched kelp

These forms display branching that is the result of a regular splitting of the meristem (an area of concentrated cell division). *Egregia*, whose irregular branches arise as outgrowths of existing branches and not from regular splitting, is a winged kelp (see p. 82). In some instances the branching is expanded (*Macrocystis, Lessoniopsis*); in others it is greatly compacted (*Nereocystis, Postelsia, Dictyoneurum*). Generally, these forms are the most spectacular of kelp by virtue of their size or habitat.

Dictyoneurum/Dictyoneuropsis

Dictyoneurum/Dictyoneuropsis (Greek=net of cords) are each represented by one species only: *Dictyoneurum californicum* and *Dictyoneuropsis reticulata*. Both species consist of a few to many blades arising from a shoe-like holdfast. The blades, sometimes reaching a few metres in length, have a net-like arrangement of veins and display splits at their bases. *Dictyoneuropsis* has a midrib and *Dictyoneurum* does not. Plants are reported to be distributed from BC to central California in the low intertidal and subtidal regions. However, a recent survey failed to verify their presence in Oregon.

Steve Fain, in a pioneering molecular study conducted at Simon Fraser University, was unable to genetically distinguish the genera *Dictyoneurum* and *Dictyoneuropsis*. Subsequent studies at Simon Fraser University, using DNA comparisons, have shown kelp to be more genetically similar than their morphological distinctiveness would suggest.

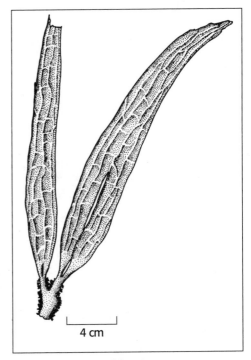

Dictyoneurum californicum.
Note shoe-like holdfast.

4 cm

Nereocystis
Bull kelp

Photo 39, p. 32

Nereocystis (Greek=mermaid's bladder) has only one species, *N. luetkeana*. Plants consist of a long stipe (up to 36 m/118') attached to the ocean floor by a holdfast composed of numerous haptera (finger-like projections) and terminated above, on the ocean surface, by a single float from which a cluster of tightly branched smooth blades arise. The blades are long (up to 4 m/13') and narrow (usually less than 20 cm/8"

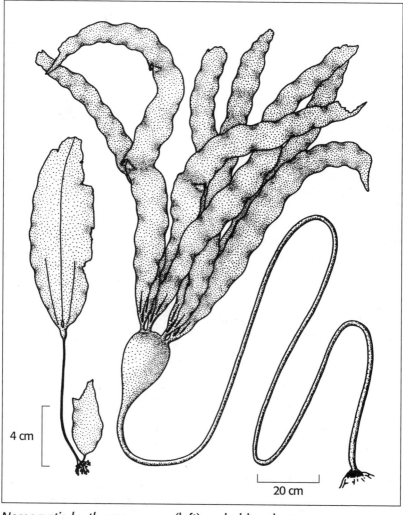

4 cm

20 cm

Nereocystis leutkeana, young (left) and older plants.

wide). Overall, this species reminds me a very large gothic brown onion of extraterrestrial origin. This form, commonly referred to as bull kelp, is attached subtidally but forms surface canopies throughout its distribution from Alaska to central California.

Several features make this species unique. It is an annual plant, although some members persist into a second year. This means that the plant achieves its significant length in one growing season (most growth occurs between March and September). To reach the maximum stipe length of 36 m (118'), the plant must grow an average of 17 cm (7") per day over the approximate 210-day period. *Nereocystis* has a logistic problem in completing its life cycle. The spores are produced on the blades at the ocean surface, often several metres above the ocean floor, but a critical concentration of spores is required near where the parent plant is successfully established to assure re-occupation of this optimal space once the annual plant is lost. So the sorus (spore patch) drops from the blade and delivers its concentrated spores to the bottom before releasing the spores. This is the only kelp to release spore patches.

Ronald E. Foreman, in pursuit of his PhD (University of California, Berkeley, 1970), discovered that the float, which may have a volume of up to 3 L (3 qts), has carbon monoxide, an infamous poison, as one of its buoyancy gases. Foreman (University of BC) is currently studying the commercial cultivation of red seaweeds. Some years ago, I had the opportunity of testing an herbivore's ability to detect the kelp-packaged carbon monoxide. While teaching a seaweed course for the University of Alaska, I shared an apartment complex with Bo, a circus elephant. I presented Bo with an entire fresh bull kelp. His response was to yank the plant from my hands (poor table manners) and eat the blades. Then, to my surprise, he stomped on the float, releasing the gas before he ate it. Does this behaviour suggest elephants once lived in association with kelp and learned to avoid the poisonous gas?

Postelsia
Sea palm
Photo 40, p. 33

Postelsia (after A. Postels, an Estonian naturalist employed by the Czar of Russia to explore our coasts), represented by one species, *P. palmaeformis*, is a magnificent, surf-loving kelp. This annual plant consists of a hollow stipe (up to 60 cm/24" long) severely attached to rock by a multi-fingered holdfast and terminated above by numerous blades arising from a tight branch system. The blades, which may achieve 40 cm (16") in length, have well-defined grooves running their length. This species, commonly referred to as sea palm, is restricted to the upper intertidal region on fully wave-exposed shores from BC to Monterey, California.

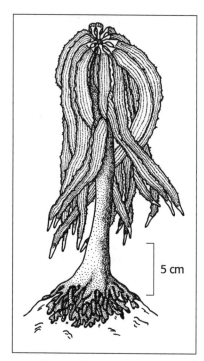

Postelsia palmaeformis.

5 cm

Plants inhabiting wave-exposed situations must adapt to considerable stress. Mark Denny has found that this habitat has water velocities possibly exceeding 14 m per second (31 mph) and accelerations that may pull in excess of 41 times the pull of gravity. *Postelsia* appears unique in adapting its spore release to the tumultuous conditions of the surf zone. Spores are released into the grooves of the blades while the plant is exposed at low tide. Before being submerged, the plant is subjected to spray, which washes the spores from the grooves onto the underlying substrate. There the spores quickly adhere before the next onslaught of surf.

Lessoniopsis
Strap kelp
Photo 41, p. 33

Lessoniopsis (after Lesson, a German phycologist), represented by one species, *L. littoralis*, is a woody, stout kelp occupying the low intertidal region in wave-exposed areas. The profusely branched plants may reach 2 m (6¹/₂') in length and consist of up to 500 terminal branches. Overall, the plant consists of an openly branched stipe system, each branch being terminated by a midrib-bearing blade. Pairs of sporophylls (special spore-bearing blades) are located at the base of older blades.

The highly convoluted stipe base may exceed 20 cm (8") in diameter and is as woody as it can be without being wood (true wood consists of special cells, xylem, which contain lignin). Japanese interests have taken out patents on the process of using kelp tissue to manufacture paper pulp. One advantage would be the absence of lignin, which must be removed from wood in the papermaking process.

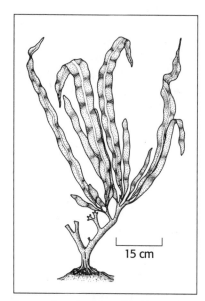

15 cm

Lessoniopsis littoralis. Note pairs of sporophylls at the bases of the ribbed blades.

Macrocystis
Giant kelp
Photos 42, 43, p. 33

Macrocystis (Greek=large bladder), the largest of all seaweeds, is represented by two species along our shores. *Macrocystis integrifolia* is the northern species, distributed from Alaska to Monterey, where it normally inhabits the lower intertidal and upper subtidal regions in areas subjected to moderate waves. *Macrocystis pyrifera* is distributed from Monterey to Mexico with suspected populations in northern BC and Alaska. This species is restricted to the relatively deep subtidal region

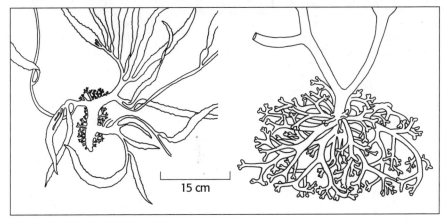

15 cm

Left: *Macrocystis integrifolia*. Right: *M. pyrifera*, showing details of holdfast. AH

in areas exposed to open ocean conditions. The two species are distinguished on the basis of their holdfasts. *Macrocystis pyrifera* has a cone-shaped holdfast composed of many haptera (finger-like projections), and *M. integrifolia* has a shoe-like holdfast with numerous haptera arising from its margin. The southern form may obtain lengths of up to 55 m (180') and the northern form may be as long as 30 m (99'). Otherwise, the plants are very similar. Both consist of numerous branches arising from a common holdfast by means of a few "Y" branchings. The first few blades arising from each branch have no floats at their bases. These are branch initials and sporophylls (special spore-producing blades). Above these are numerous vegetative blades, each with a basal float. At the top of the branch is a distinctive blade in which numerous splits may be observed, from which other blades grow.

Macrocystis is the major canopy-forming seaweed. Its expansive canopy modifies oceanic conditions such as water motion and light penetration, giving rise to a great number of habitat opportunities.

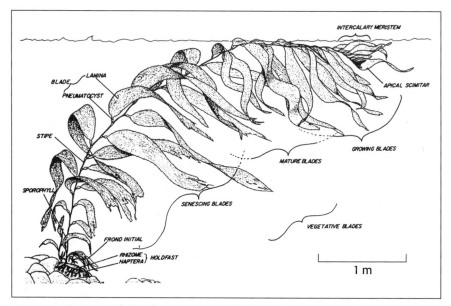

Macrocystis integrifolia, the gross morphology of a mature plant.

Mike Neushul prefaced his doctoral thesis (University of California, Los Angeles, 1959) with this long sentence by J.D. Hooker (1847), expressing this phenomenon: "So many interesting points are connected with *Macrocystis*, that a book might be instructively filled with its history, anatomy, physiology and distribution: whilst its economy, its relation to the other vegetables and to the myriads [sic] of living creatures which depend on it for food, attachment, shelter and means of transport constitute so extensive a field of research that the mind of a philosopher might shrink from the task of describing them."

Macrocystis and *Nereocystis* are the major canopy-forming species in our area. The unique habitat created by these species is missing from north Atlantic coasts, which may limit species diversity of those shores.

Macrocystis is harvested extensively in southern California, and concern for the survival of kelp beds has led to numerous researches—not only because of its economic potential. I like to consider the renaissance of seaweed field biology, including *Macrocystis* ecology, as having been started by Professor Mike Neushul (University of California, Santa Barbara). Neushul was a pioneer in applying SCUBA in the academic pursuit of marine botany. In fact, his whole career was marked by discoveries of unique approaches to ocean studies. Before cancer cut his career short when he was only fifty-nine, Neushul trained twenty-one marine botany PhDs, many of whom are still active on our coast.

Red Seaweeds

Red seaweeds make up the majority of seaweed species. Up to 5,500 red seaweed species are recognized worldwide. Of the approximately 600 seaweed species noted for our survey area, approximately 400 are red seaweeds. These seaweeds vary morphologically from simple filaments to intricate, fleshy plants. Their most elaborate form is expressed as dorsal-ventrality, or having a distinguishable front and back.

"Red" algae can appear greenish, yellowish, brownish, red (including pink) or purple. Red, their basic colour, is the result of the dominant pigment phycoerythrin (Greek=algal red) and associated phycocyanin (Greek=algal blue). When this pigment complex is destroyed or diminished by stressful conditions of intense light, high temperature and/or low nutrients, the masked pigments responsible for green and brownish coloration are expressed, thus giving the seaweed an off-colour. At first this may cause confusion in identifying a seaweed, but once you become familiar with the basic forms of red, green and brown seaweeds, colour will not be so important. Off-colour reds often have a little of their true colour near the point of attachment. Also, due to their unique intercellular connections, red seaweeds have a rubbery elasticity not found in other seaweeds. You can feel this by gently tugging a fleshy form.

Red seaweeds store a starch very similar to that found in green plants—a starch that is digestible by humans. These plants have cellulose as a major cell wall component, like all multicellular plants. Uniquely, they often have the plant gums agar and carrageenan, which are employed commercially as emulsifiers. The cell walls of pink forms of red seaweeds, known as corallines, are penetrated with calcium and magnesium carbonate, which makes the plant stone-like. The red algae lack any motile stage. This should complicate the bringing together of male and female gametes for reproduction, but they seem quite successful.

Hard crusts (calcified)

The red seaweeds known as coralline algae toughen their bodies by impregnating themselves with calcium and magnesium carbonate, resulting in a tooth-like hardness. These seaweeds take on two basic forms: crustose corallines (crusts) and geniculate corallines (erect, branched plants). The many species of crustose coralline algae are difficult to distinguish in the field and, because of their rock-hardness, are a nuisance to identify in the laboratory. (Their carbonates must be dissolved with weak acids, like vinegar, before they can be sectioned and examined microscopically.) Their colour is usually pink but may tend toward purple. Hard crusts may grow on other seaweeds or hard substrate, including rock and hard parts of various animals. Robert Steneck (University of Maine) has studied our local crusts extensively. A general conclusion of his studies is that in our area hard crusts are the major occupiers of primary substrate through the low intertidal and subtidal regions, where drying is unlikely (pools, shaded areas, under kelp canopies).

Pseudolithophyllum
Rock crust

Photos 44, 45, p.33

Pseudolithophyllum (Greek=false rock leaf) is used here to represent the five genera and nine species of crustose coralline algae growing on hard surfaces and variously distributed throughout the intertidal and subtidal regions, from Alaska to Mexico. *Pseudolithophyllum* is the most common local genus with species dominating the mid- to high intertidal region (*P. neofarlowii*) and the low intertidal region (*P. muricatum*). *Pseudolithophyllum neofarlowii* is the only crustose coralline found in the upper intertidal zone, where it occupies steep, shaded surfaces. It forms vast pink to purple to chalky-appearing crusts, which are covered with wart-like projections. *Pseudolithophyllum muricatum* is a major competitor for space in the lower intertidal zone by virtue of its ability to grow over other crusts, some fleshy seaweeds and even sessile animals. The plant is a smooth, undulating circular patch with a leafy (free from the substrate) white margin. In contrast, some crustose species have highly sculpted surfaces (e.g. *Lithothamnion*).

Due to their hardness, these crusts were thought to be protected from grazers. Robert Steneck, however, has demonstrated that these plants are frequently grazed, which is necessary to protect them from overgrowth by other sessile organisms. During this grazing/cleaning, the crust surface is damaged but the hardness and thickness of the crust minimize the impact.

Melobesia
Seagrass crust

Photo 46, p. 34

Melobesia (Greek=daughter of Oceanus) is represented by two crustose corallines growing on other fleshy marine plants. These appear as simple pink discs up to 5 mm (3/16") in diameter. *Melobesia marginata* grows on the red seaweeds *Ahnfeltia* (see p. 122) and *Osmundea* (p. 119), and *M. mediocris* grows on the seagrasses *Phyllospadix* (p. 43) and

Zostera (p. 42). These minute plants are found on their hosts in the low intertidal zone from BC to Mexico.

Mesophyllum
Coralline crust

Photo 47, p. 34

Mesophyllum (Greek=inter-mediate leaf), consisting of two local species, is included here to represent several crustose corallines that usual-ly grow on non-crustose branched or geniculate corallines. These pink to lavender plants are more or less circular, up to 3 cm (1 1/4") in diameter. The cen-tral portion of the plant is attached to the host, leaving the wavy margins free. Many of these seaweeds are reminis-cent of wood-rotting fungi by virtue of their circular outline and the circular pattern of pores and lines. *Mesophyllum* is found on its hosts in the mid- and low intertidal and subtidal regions from Alaska to southern California.

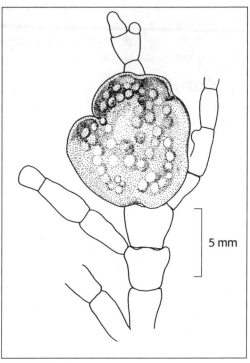

5 mm

Mesophyllum conchatum on *Calliarthron.*
AH

Branched and hardened (calcified)

The calcified and branched red seaweeds (known as geniculate—or jointed—corallines) represent one of the most curious plant forms. The calcification provides some protection from grazers while presenting the plant with two unique problems. To reproduce successfully and disperse spores, the armoured plant needs to provide access to its internal recesses for the release of spores and sperm, and to permit access to eggs. The second problem, amplified for intertidal plants, is one of flexibility. These plants cannot exist in the presence of strong water motion, as rigid, brittle systems. They need to live with the strong drag forces they experience. This problem is solved in much the same way as knights in armour retained their flexibility—with soft joints. Close examination of a geniculate coralline reveals a system of hardened segments alternating with flexible joints, much like beads on a string. The underlying cells show a similar alternation of hinge-like and non-hinge-like cells.

The geniculate corallines frequently contribute to the shell sand of pocket beaches along our western coasts. Careful inspection of the sand reveals white bleached calcified segments, bead-like in appearance.

Following are descriptions of three genera representing seventeen common local species.

Corallina
Coral seaweed

Photos 48, 49, p. 34

Corallina (named for their resemblance to coral animals) consists of four species that may be distinguished from most other species by the conceptacles (pores for the release of reproductive spores and sperm) located on the tips of branches. These appear as miniature white-rimmed vases. The purplish-pink plants consist of mostly rounded calcified segments 1–2 mm (^{1}/32–^{1}/16") in diameter and can reach 10 cm (4") in length. *Corallina officinalis* var. *chilensis* and *C. vancouveriensis* are two conspicuous species. *Corallina vancouveriensis* is found in the low to mid-intertidal region. This species has small calcified segments and tends to be branched in the round, whereas *C. officinalis* var. *chilensis* is found in the lower intertidal region, has larger calcified segments and a distinct flattened, feather-like branching pattern. *Corallina* may be found in high tide pools, in shaded mid-intertidal regions and in the low intertidal region from Alaska to Mexico.

Corallina officinalis L. was the name Carolus Linnaeus applied to what he thought was an animal coral in 1758. Almost a hundred years passed before it

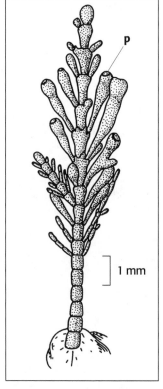

Corallina vancouveriensis. Note reproductive pores (p) on the ends of some branches.

was recognized that *Corallina* and related corallines were plants.

Corallina vancouveriensis was discovered by K. Yendo of Japan while visiting the Minnesota Seaside Station on Vancouver Island. This station, the first marine station in the Pacific Northwest, was established by Josephine Tilden, University of Minnesota, in 1900. For two years, Professor Tilden produced the journal *Postelsia*, in which were published intriguing fireside chats with visiting scientists. Gayle Hansen (Hatfield Marine Science Centre, Newport, Oregon), a seaweed taxonomist emphasizing conservation biology, has dedicated considerable effort to understanding this phyco-pioneer and is writing an expanded biography of Josephine Tilden.

Bossiella
Coral leaf

Photos 50, 51, p. 35

Bossiella (after Bosse, a French phycologist) represents five species of markedly flattened plants bearing pores on one surface. These pores are lined up parallel to a more or less conspicuous midrib running the length of the calcified segments. Plants are decidedly pink, reaching 15 cm (6") long. *Bossiella* is found in the low intertidal zone from Alaska to Mexico.

Calliarthron
Bead coral

Photo 52, p. 35

Calliarthron (Greek=beautiful joint) consists of two species of robust plants distinguished from most other articulate corallines by the pores it bears on the margins and flattened surfaces of the cylindrical to slightly flattened branches.

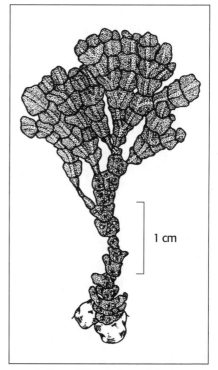

Bossiella californica. Note reproductive pores on the flat surfaces.

Plants are up to 25 cm (10") long, forming dense, often expansive wiry brilliant pink turf in the lower intertidal region. *Calliarthron* is common from Alaska to Mexico.

Brenda Konar and Mike Foster (Moss Landing Marine Laboratories, Moss Landing, California) noted that *Calliarthron* was the dominant subtidal seaweed in Carmel Bay, California, occupying up to 40 percent of available surfaces. They followed the development of this near-monoculture on cleared spaces. New crusts, growing 0.3–0.6 mm per month, gave rise to upright branches after sixteen weeks. The branches elongated 0.01–0.5 mm in length per month. After forty-eight weeks the cleared area and adjacent, undisturbed areas were indistinguishable. The clearings used in this study were achieved with a pneumatic chisel. Ecologists will go to almost any extreme to discover marine "truths," as noted in the ditty "Perturbation," which introduced a lecture by Robert Paine at Simon Fraser University in 1983.

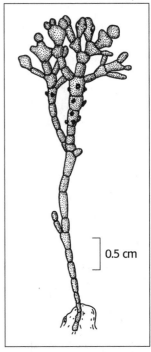

Calliarthron tuberculosum. Note the reproductive pores on sides and fronts of the segments.

With pickax and dynamite,
Like surgeon's tools,
He delicately probed,
To unravel ecology's rules.
No Paine was too great!
With flame-thrower and cage,
Like an academic soldier,
He mightily extracted
The diversity of each teeming boulder.
No Paine was too great!
All plants and animals
Did his studies involve.
For it was his to—
Diversity, foodwebs and stability resolve.
No Paine was too great!

Soft crusts

There are many red soft-crusted species on our shores. To determine that you are looking at a red crust and not a crustose lichen or brown crust, moisten the crust: if it appears red, orange red or purplish it is a red seaweed. Crusts have a low profile, adhering tightly to their substrate, with no significant upright extensions. Once they become established they may dominate their area for decades. Physically, they are difficult to graze and their comparatively low surface area to volume ratio gives them resistance to atmospheric influences (drying, freezing, UV) during periods of exposure. In some instances, crusts are known to alternate with more morphologically elaborate phases. This type of life cycle allows the plant to exist in different environments and/or during different seasons. For example, a persistent crust may re-inoculate, through spore production, a more elaborate and delicate seasonal phase.

While probing about in the intertidal region, note the abundance of various crusts. Little space is unoccupied. The ecology of the soft crusts (red, brown, green and lichen) has been an area of interest for Megan Dethier, who has shown the non-specialist how to distinguish some forms in the field and most forms with the aid of a microscope.

Mastocarpus
Turkish washcloth (crust phase)
Photos 53, 54, p. 35

Mastocarpus (Greek=whip-like body) is the crust phase of *Mastocarpus*, the simple-blade-with-regular-bumps (see p. 110) phase. This crust is slightly spongy. If it indents under pressure from your fingernail it is most likely *Mastocarpus*. If not, it is most likely *Hildenbrandia* (see below). *Mastocarpus* crusts vary in size from a few cm to 30 cm (12") across. They are deep red (sometimes appearing near black) and occupy the high to low intertidal regions, being most common lower down, from Alaska to Mexico.

John West (University of California, Berkeley) discovered, through culture studies, the connection of the two phases of *Mastocarpus* in a single life cycle. He observed spores from the crust phase giving rise to the bladed sexual phase, which through production of eggs and sperm gave rise to spores, which in turn gave rise to the crusts. West has made similar observations on numerous red algae, greatly enhancing our understanding of these seaweeds. West's prominence as a phycologist is unquestioned in his guild. However, for one brief Warholean moment, this fame was overshadowed thanks to North American chocoholics. In 1982 he offered a course on the Biology of Chocolate at the University of California, Berkeley, and according to *Time* magazine he had over 800 applicants.

Scientists have suspected that the crust phase persists for a long time, and Robert Paine had the patience to document it. He determined a growth rate of 4 percent per year, so the measured crusts were twenty-five to eighty-seven years old. Paine has made a phenomenal impact on our understanding of seaweed biology, particularly when you consider he is a professor of zoology. He and his students have studied aspects of seaweed dispersal, population structure and plant/herbivore interactions—I suspect through the eyes of invertebrates.

Over half of the lower-positioned crusts house the unicellular *Codiolum*, the spore-producing phase of many green seaweeds (see Unicellular forms, p. 55).

Hildenbrandia
Rusty rock

Hildenbrandia (named after a person about whom I know nothing!) is represented by three species found in the high to low intertidal regions. These plants are expansive, covering large areas of rock. They look like *Mastocarpus*, but most may be distinguished by not being spongy and not indenting under the pressure of your fingernail. Also, they often appear rusty or bright red in colour and not as dark as *Mastocarpus*. They tend to be more abundant higher up than *Mastocarpus* and dominate high tide pools. These forms are distributed from Alaska to Mexico. Closely related forms are found in fresh water.

Parasitic plants

Photo 55, p. 36

The parasitic red seaweeds parasitize other red seaweeds. They are all small (usually projecting a few mm above the host's surface) and are either white or pale red/pink. Locally, several genera of parasitic red seaweeds are recognized. These parasites and their hosts are variously distributed from Alaska to Mexico. They may be distinguished by their hosts. Ten of the potential hosts (those described in this text) and their parasites are listed below.

HOST	PARASITE	PARASITE MORPHOLOGY
Cryptopleura	*Gonimophyllum*	rosettes of small pale blades
Polyneura	*Asterocolax*	clusters of needle-like branches
Gracilariopsis	*Gracilariophila*	warty cushions
Plocamium	*Plocamiocolax*	branches arranged as teeth on a comb
Palmaria	*Rhodymeniocolax*	branches arranged as grapes in a bunch
Polysiphonia & *Pterosiphonia*	*Leachiella*	cushion-like
Odonthalia & *Neorhodomela*	*Harveyella*	globose pustules
Osmundea	*Janczewskia*	warty cushions

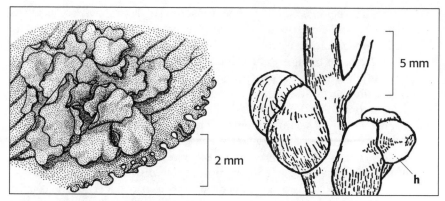

Left: *Gonimophyllum skottsbergii* on part of a *Cryptopleura* blade.
Right: *Harveyella mirabilis* (h) on a *Neorhodomela* branch. AH/TBW

Our understanding of red parasitic algae and their relationships to their hosts is largely the result of studies by Linda J. Goff (University of California, Santa Cruz) and her associates. Earlier, these parasites were considered galls, or globs of amorphic tissue, originating from the "host" plant. However, developmental and physiological studies, such as those initiated by Linda Goff in her PhD studies with Kathleen Cole (University of BC, 1975) have revealed intriguing host/parasite relationships. The parasites accommodate the transitory nature of their living substrate by completing their entire life cycle in as little as eight weeks. Goff and others consider this parasitic relationship advanced because there is little destruction to host tissue, the host tissue changes to accommodate the parasite, and the parasite is usually restricted to one host species.

Most recently Goff and associates have determined that many of these red parasites have evolved from their hosts. This reminds me of the story about a zoologist who walks into a store with a parrot on his shoulder and the clerk asks, "Where did you get that?" and the parrot responds, "It started as a wart on my butt."

Filamentous

This seemingly simple body plan presents itself in a bewildering array in the red seaweeds. Filamentous plants may be one cell wide or several, unbranched or branched in a great many patterns, erect or creeping, and consist of uniform or differentiated cell shapes. In spite of their apparent simplicity, they have the same body parts and functions found in the most sophisticated fleshy red seaweed.

Fleshy red seaweeds often have free-living filamentous phases that may allow them to persist through adverse times. For example, *Porphyra* (see p. 114), a fleshy red seaweed often found in the high intertidal regions, has a filamentous phase during which it exists inside shells. This phase may survive better over summer than the fleshy phase and re-establish the fleshy phase in the fall, when conditions are more benign. Jane Lubchenco and John Cubit (University of Oregon) noted that the presence of fleshy stages of some intertidal seaweeds in the fall-to-spring period coincided with diminished herbivore activity. For the remainder of the year, only the filamentous phase persisted. When scientists removed the herbivores during the summer period (the time of heavy herbivory), the fleshy stages were established. This study nicely demonstrates an advantage of having a heteromorphic life cycle (an alternation of two distinct morphologies).

There is no question filamentous forms can successfully coexist with the physically domineering seaweeds such as kelp. Often they live attached to these larger forms, thus avoiding being totally shaded or deprived of nutrients. Not only can they hang on to their major competitors, but also there is growing evidence they can tap their living substrate for energy-rich compounds.

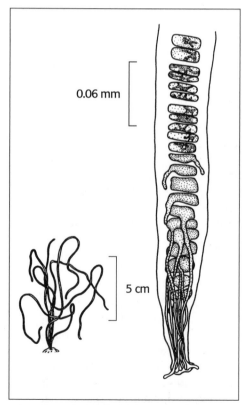

0.06 mm

5 cm

Bangia and detail (right) of cellular arrangement.

Bangia
Black sea hair

Bangia (after N.H. Bang, a Danish botanist) consists of one or perhaps two marine species. Evidence for two species is based on chromosome counts (some plants have three chromosomes and others have four). Plants are unbranched, one or many cells wide and up to 10 cm (4") long. They appear as rusty-black hair plastered on rocks and pilings at the high tide level from Alaska to Mexico.

One species of *Bangia* lives in fresh water. C. den Hartog (Holland) conducted a comparative study of marine and freshwater *Bangia*. He could adapt the freshwater species to the marine environment and vice versa, and concluded they may be the same species. He also noted that *Bangia* would be found only in freshwater or marine situations and not in intermediate estuarine conditions. This could indicate *Bangia* is intolerant of fluctuating salinity.

Antithamnionella
Red sea skein

Antithamnionella (Greek=opposite branching little bush) is represented by three or four species of delicate, profusely branched plants consisting of filaments one cell wide. The terminal branching system consists of pairs of opposite branchlets arising from the penultimate branches. To see this feature may require a magnifying glass. Plants are up to 10 cm (4") long and usually densely clustered. When observed at

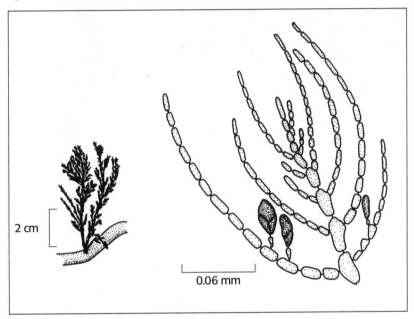

Antithamnionella pacifica and detail (right) of opposite branching.

low tide they appear as a dark red glob. *Antithamnionella* is distributed from Alaska to Mexico, where it is found in the low intertidal region. From Alaska to central California it may be observed on the stipes of the bull kelp (*Nereocystis*, p. 90).

James Markham (University of California, Santa Barbara) studied the distribution of *Antithamnionella* along the stipes of *Nereocystis* as part of his Master of Science study (University of Washington, 1963). He noted vegetative plants nearest the surface, followed lower by female plants, then by spore-producing plants (tetrasporophytes, see figure on p. 18), an arrangement that is mirror-imaged as one proceeds down the stipe. This arrangement of life cycle stages was observed on long and short stipes, suggesting light alone was not the deciding factor, as light presents a regular depth gradient. The same pattern was produced on man-made plants (garden hose), eliminating the natural substrate as a deciding factor. What possible explanations are there for these phenomena?

Callithamnion
Beauty bush

Photo 56, p. 36

Callithamnion (Greek=lovely little bush) is represented by up to six species, three of which are encountered intertidally. In the field, their most distinguishing feature is their fuzzy, soft appearance. They usually appear brownish. This group has a staggered arrangement of branchlets (you need a magnifying glass to see this), in contrast to *Antithamnionella*, where the branchlets are in pairs. The branches are mostly one cell wide. These plants, up to 40 cm (16") long but usually less than 10 cm (4"), are distributed from Alaska to southern California anywhere in the mid- to low intertidal and subtidal regions.

3 cm

Callithamnion pikeanum.

Ceramium
Staghorn felt

Photo 57, p. 36

Ceramium (Greek=a vessel) includes approximately ten species, of which about half are rare or exclusively subtidal. The basic axis of these

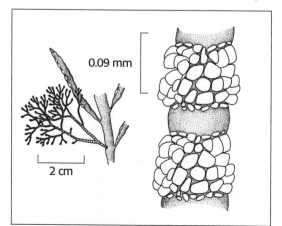

0.09 mm

2 cm

plants is one cell wide, regularly interrupted with narrow clusters of smaller cells. This arrangement results in a banding pattern that is visible to the naked eye. The branch tips are frequently pinched in, looking like crab claws. Most

Ceramium californicum on a coarser red seaweed (left). Detail of alternating cell types (right)

species are small (less than 2 cm/3/4" long) but some reach 10 cm (4") in length. An easy place to find *Ceramium* is on the green seaweed *Codium fragile* (see p. 53), which hosts two species. *Ceramium* may be found in the mid- and low intertidal and subtidal regions on rock, other plants, animals or soft bottom from Alaska to Mexico.

Polysiphonia
Polly

Polysiphonia (Greek=many tubes) consists of up to ten species. Plants of this genus are radially branched, resulting in a fluffy appearance when seen in the water or exposed on the beach. The slender branches (less than 0.2 mm in diameter) are cylindrical and many cells thick. The outer cells surrounding the branch are polysiphonous, or arranged much like bundles of same-length straws stacked bundle-on-bundle. Plants are up to 25 cm (10") long and brownish to sharp red. *Polysiphonia* is frequently encountered in the mid- and low intertidal and subtidal regions from Alaska to Mexico in sheltered and wave-exposed regions.

Worldwide, *Polysiphonia*

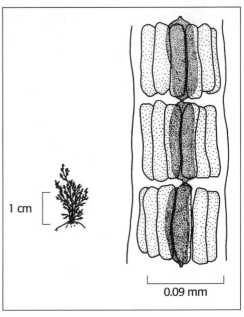

Polysiphonia hendryi and detail (right) of the polysiphonous condition.

consists of about 150 species. As with all complex taxonomic groupings, one must question whether there is such a large number of species. Are they genetically unique, incapable of sexual crossing, or are they capable of interbreeding? The noted morphological differences may result from a similar genotype that has either genetically differentiated into an ecotype or that responds to differential environmental conditions by modifying its morphology (morphological plasticity). Jan Rueness, a Norwegian phycologist, addressed this question by attempting to cross two species, one from Norway and one from the Gulf of Mexico. Both species were interfertile but their subsequent ability to produce spores was limited. Rueness concluded these two entities were varieties within one species.

Pterosiphonia
Black tassel

Pterosiphonia (Greek=winged tubes) consists of six local species whose cell arrangement is similar to that described for *Polysiphonia*. *Pterosiphonia* may be distinguished from *Polysiphonia* by the flattened arrangement of the branches. Plants are up to 25 cm (10") long, ranging from blackish to red, and are found in the mid-intertidal to subtidal regions. Plants are common in areas subjected to waves from Alaska to Mexico.

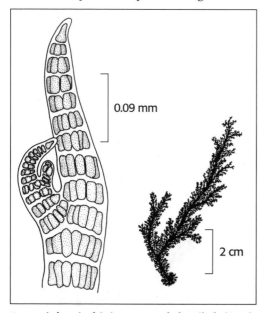

0.09 mm

2 cm

Pterosiphonia bipinnata and detail (left) of the polysiphonous condition.

Solid or hollow cylinders

The cylindrical morphology, found amongst brown (*Scytosiphon*, p. 70) and green (*Enteromorpha*, p. 52) seaweeds, is shared with some land plants (cacti). This unembellished form results in a reduced surface area per unit plant volume, a feature that may benefit intertidal plants subjected to drying during low tides or, in the case of cacti, exposed to dry desert air. The hollow forms are often buoyed up in the water by captured air that may give them an advantage over their low-lying neighbours in competing for light.

Halosaccion
Dead man's fingers

Photo 58, p. 36

Halosaccion (Greek=sea sac) is represented locally by one species, *H. glandiforme*, found in the mid- to low intertidal regions from Alaska to southern California. Sacs ranging in colour from yellowish to deep red and standing up to 20 cm (8") tall are found in dense clusters. They look like fingers of a rubber glove, hence the common name.

When one of these sacs is squeezed, fine jets of water shoot into the air much like water jets from a fireboat. To observe this, hold the sac

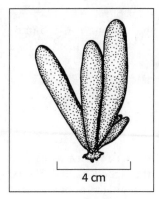

Halosaccion landiforme.

near your eyes and squeeze. Steve Vogel (Duke University) explored the hydromechanics of these pores and concluded they protect the plant from desiccation by allowing the re-entry of water following a period of emersion. Professor Vogel has been largely responsible for introducing the discipline of biomechanics to students of marine organisms. While Steve Vogel was teaching biomechanics, one of his students, Mimi Koehl (University of California, Berkeley), introduced him to the real world of applying this discipline to field situations. This two-way interaction between teacher and student accelerates the accumulation of knowledge, enriches all participants, and inspired the dedication of this book.

Nemalion
Rubber threads

Photo 59, p. 37

Nemalion (Greek=a thread) is represented locally by one species, *N. helminthoides*, which is found in the mid- to lower intertidal regions from Alaska to Mexico. Deep red, cylindrical plants up to 5 mm ($^3/_{16}$") in diameter and 40 cm (16") long (rarely 100 cm), are found in clusters, usually in the spring. This distinctive worm-like morphology is echoed in the species name: *Nemalion* (nematode) *helminthoides* (like a parasitic worm).

Blades with regular bumps

Many bladed red seaweeds produce bumps in association with reproduction. These are visible to the naked eye, but they are quite small and usually present only seasonally. The plants

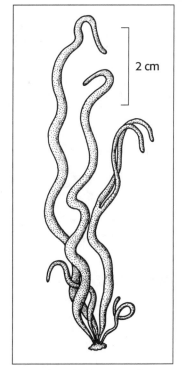

Nemalion helminthoides.

noted in this section have conspicuous, persistent bumps (papillations) that are not necessarily associated with reproduction. These bumps may benefit the plants by causing turbulence close by, to introduce nutrient-rich sea water to the blade surface.

Chondracanthus
Turkish towel

Photo 60, p. 37

Chondracanthus (Greek=spiny *Chondrus*—another red seaweed) is a complex of five to eight species of simple unbranched bladed plants, which may be irregularly divided, bearing regular bumps. These bumps are vegetative features and not necessarily associated with reproduction. The blades are tough and their colour ranges from red to yellow. Plants may approach 80 cm (32") in length but usually are less than 30 cm (12") long. Within this assemblage, there are less conspicuous branched forms that are represented below under *Mastocarpus*. Representatives of *Chondracanthus* are found in the low intertidal and subtidal regions in wave exposed and sheltered waters from Alaska to Mexico. Until recently, *Chondracanthus* was known as *Gigartina*.

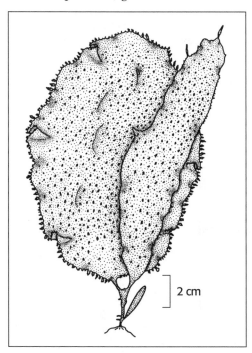

Chondracanthus exasperatus.

The larger forms, up to 40 cm (16") long, are called Turkish towel because of their rough texture. They are used in bathing, much as a loofah (the vascular system of dried squash), to remove dead skin.

Mastocarpus
Turkish washcloth (erect phase)

Photos 61, 62, p. 37

Mastocarpus (Greek=whip-like body) consists of two species, formerly considered to be *Chondracanthus* species. These species have poorly defined branching that approaches being dichotomous ("Y"-shaped), with the terminal branches curled as you would curl your tongue. Plants are usually less than 15 cm (6") long. *Mastocarpus* is differentiated from *Chondracanthus* on the basis of its life cycle. Whereas *Chondracanthus* has alternating isomorphic generations (the spore-bearing and sexual, gamete-bearing phases are morphologically identical), *Mastocarpus* has alternating heteromorphic generations where the spore-producing phase is a crust (see p. 101) and the gamete-producing phase is a papillate (with bumps) blade. *Mastocarpus* is found from Alaska to Mexico in the mid- to low intertidal regions of wave-exposed and moderately wave-sheltered areas, often associated with *Fucus*.

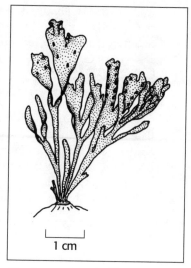

1 cm

Mastocarpus papillatus.

Blades with ribs

Ribs and veins are common features in many seaweed groups. The following examples represent the range of venation you are most likely to encounter. Ribs, which are most conspicuous and protrude above the blade surface, are thought to serve as a strengthening mechanism for otherwise delicate plants. Veins, which are less conspicuous, are often lines created by the distribution of reproductive structures. Veins and ribs may serve as primitive vascular systems, facilitating translocation (the movement of nutritive compounds through the plant).

Erythrophyllum
Red sea leaf

Erythrophyllum (Greek=red leaf) is represented by two species of spectacular plants that are found in the low intertidal region of wave-exposed shores. The rich, red blades are narrow, up to 40 cm (16") long, bearing conspicuous ribs running their length. The margins of the blades are often notched to the rib. No excursion to the open coast is complete without seeing these beautiful plants. Typically they are located in the lower intertidal region on the

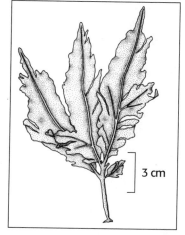

3 cm

Erythrophyllum delesserioides.

walls of surge channels, where the force of waves is amplified, their rich fronds swirling in the white, frothy surge. These plants are distributed from BC to central California.

Cryptopleura

Photo 63, p. 37

Cryptopleura (Greek=hidden ribs) is represented by four or so species usually having visible veins radiating out along deeply divided blades. The veins branch repeatedly from the base of the blade, becoming less conspicuous toward the branch tip. The blades, up to 40 cm (16") long, often have frilly margins. Plants are found in the lower intertidal and subtidal regions from Alaska to Mexico.

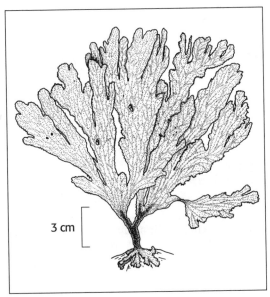

3 cm

Cryptopleura ruprechtiana. Note open network of veins.

Cryptopleura is one of a series of similar forms of often confusing genera, including *Hymenena* and, *Botryoglossum*. Eugene Kozloff (Friday Harbor Laboratories, University of Washington), a master of local seashore life as well as opera and baroque music, noted: "A novice need not feel badly about remaining noncommittal" (not sure about which genus you are dealing with).

Look your specimen over carefully for little cabbage-like outgrowths. These may be the parasitic red seaweed *Gonimophyllum* (see p. 102).

Polyneura
Crisscross network

Polyneura (Greek=many nerves) consists of one local species, *P. latissima*. This is a distinct species that may be easily distinguished from other frilly-bladed reds by having a net-like system of veins. These veins branch and fuse back into each other, producing the net-like appearance. The blades may reach 30 cm (12") in length but usually are less than 15 cm (6") long. *Polyneura* ranges in colour from pale red to pinkish. These plants are most common early in the year, when they

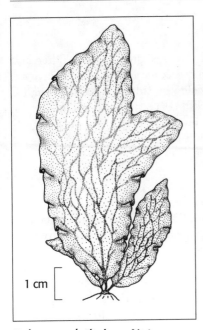

Polyneura latissima. Note closed network of veins.

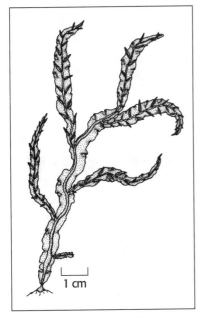

Delesseria decipiens. Note distribution of bladelets on only one side of the blade.

are found in the mid- to low intertidal and subtidal regions of moderately wave-exposed to sheltered shores. They are distributed from BC to Mexico.

Delesseria
Winged rib

Delesseria (after Baron Delessert, a French naturalist) is represented locally by one species, *D. decipiens.* *Delesseria* is a frilly, delicate, branched and bladed seaweed that has a conspicuous midrib. One side of the blade has little bladelets that arise along the midrib. The resulting morphology is referred to as dorsal-ventrality (having a front and back). These plants are often found hanging on rock walls in loose clusters. The pale red blades may reach 50 cm (20") in length. *Delesseria* is usually encountered in the spring in the low intertidal region from Alaska to central California.

Our understanding of this morphologically complex genus and related forms is the result of Robert Scagel's doctoral studies (University of California, Berkeley, 1953), conducted under the tutelage of George Papenfuss.

Delesseria is a particularly fragile plant, which if collected must be handled with care. If abused by being allowed to warm up, it will bleed its red pigments, which belong to a unique group of water-soluble pigments called phycobilins. Most photosynthetic pigments are soluble in organic solvents, not water.

Simple blades

Simple-bladed forms are superficially very similar to each other and final identification usually requires microscopic examination. However, if you were to lay the various species side by side, you would note differences that could direct you toward a correct identification. Overall the size of these forms ranges from a few centimetres to 1 m (3').

Porphyra
Purple laver

Photo 64, p. 38

Porphyra (Greek=purple) is represented by a bewildering twenty-two species in our area. *Smithora naiadum* (Seagrass laver), which grows on seagrasses, is indistinguishable in the field from *Porphyra*. Generally these species are distinguished by a combination of macroscopic and microscopic detail. The blades, which may measure up to 1 m (3') in their greatest dimension, are one or two cells thick, making them almost transparent. Their colour can vary from whitish-yellow to brownish to red and purple. They grow attached to other seaweeds, and almost anything else that does not move. Mostly they occur in the mid- to high intertidal region and often are the highest conspicuous intertidal seaweeds on the beach. Representatives of *Porphyra* and *Smithora naiadum* are common from Alaska to Mexico.

Interest in *Porphyra* arises in part from its economic value. This is the stuff of nori, the tasty seaweed wrap used in sushi and other Japanese dishes. Farming of these plants constitutes the single greatest

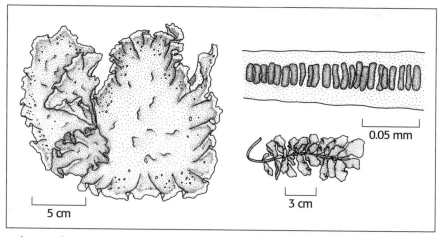

Left: *Porphyra perforata*. Top right: Cross-section of the blade, showing one-cell thickness. Bottom right: *Smithora naiadum* on leaf of *Phyllospadix*.

aquacultural enterprise currently being undertaken. This success resulted from the findings of the British phycologist Kathleen Drew, who discovered the connection between the blades of *Porphyra* and a filamentous red seaweed that inhabits the interior of shells. This discovery, which made possible a multi-billion dollar industry, moved the Japanese to build a monument to her over which a Shinto priest prays daily.

Else Conway, of the Drew lineage, joined Professor Kathleen Cole at the University of BC, where considerable interest was focused on *Porphyra*. Mike Hawkes, Tom Mumford and Sandra Lindstrom, three UBC students, have also made major contributions to our understanding of this genus. Mike Hawkes was the first to clearly demonstrate sex in this genus. This was hot! Previously, Professor Peter Dixon (University of California, Irvine) also of the Drew lineage, had declared to a large international group of eminent phycologists meeting in Spain that there was no sex in these plants. Sandra Lindstrom contributed to our understanding of the taxonomic relationships among the many species.

Palmaria
Red ribbon

Photo 65, p. 38

Palmaria (Latin=palm of a hand) is represented locally by up to five species, all being thick, not slimy and usually having deeply divided blades. The usually deep red blades may be regularly or irregularly divided. Blades may reach 40 cm (16") in length but usually are less than 20 cm (8") long. These plants are restricted to the low intertidal and subtidal regions and are variously distributed from Alaska to Mexico.

For a long time female plants of *Palmaria* were not known in the wild. John

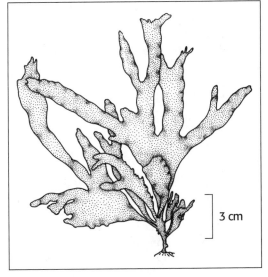

Palmaria mollis.

van den Meer of the National Research Council, Canada, discovered that female plants remained nearly microscopic, unlike the large male blades. This sexual dimorphism reduces the chances of inbreeding.

While the male is growing to a large size associated with sexual maturity, the female reaches this state quickly because of her small size. The ripe eggs are present before sib sperm are available and are fertilized by a previous male generation.

On various hostile British and Norwegian shores, sheep are put to pasture on the seaweed-infested intertidal region, and they invariably graze *Palmaria*, which they are capable of identifying without the aid of this guide.

Schizymenia
Slimy leaf

Schizymenia (Greek=split membrane) consists of two species that superficially resemble *Palmaria*. One species is infrequently encountered; the other, *S. pacifica*, is common in the wave-exposed low intertidal region from Alaska to Mexico. *Schizymenia pacifica* blades, which are clustered in dense groups, may reach 60 cm (24") in length but are usually less than 30 cm (12") long. They may be distinguished from other bladed red seaweeds by being slimy to the touch. These rich red blades are infrequently and irregularly divided.

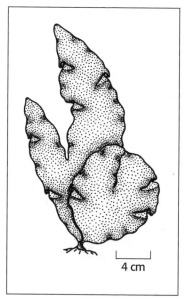

4 cm

Schizymenia pacifica.

Mazzaella
Rainbow-leaf

Photos 66, 67, p. 38

Mazzaella (after Angelo Mazza, an Italian phycologist) is locally represented by five species having two basic forms. One, *M. cornucopiae*, is restricted to wave-exposed shores from Alaska to central California. These plants consist of variously lobed blades, under 5 cm (2") long, which are densely packed into frilly stands. The other species are larger, usually longer than 20 cm (8"), and are not densely packed nor do they have a frilly appearance. *Mazzaella linearis* (wave-exposed) and *M. splendens* (wave-sheltered) are two commonly encountered species. *Mazzaella linearis* has long, narrow blades that are found in clusters, whereas *M. splendens* has broad blades that may or may not be clustered. Usually these plants are iridescent, as reflected in their older name *Iridaea* (Latin=rainbow), a feature that sets them apart from most

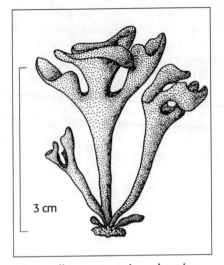

Mazzaella cornucopiae, showing rolling of the blades. AH

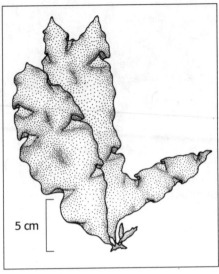

Mazzaella splendens.

other large bladed red seaweeds. Their colour varies from deep reddish purple to yellowish brown. A gentle tug reveals a striking elastic quality of non-red forms, clearly establishing them as red seaweeds. *Mazzaella* species are found in the mid- to lower intertidal regions from Alaska to Mexico.

Mazzaella has an isomorphic life cycle, where the spore-producing phase alternates with a morphologically similar gamete-producing sexual phase. These two phases can be distinguished in the field by conducting a chemical test that identifies carrageenans (economically important plant gums) unique to each phase. Robert deWreede has used this phenomenon to determine how the two phases allocate beach habitats.

Constantinea

Photo 68, p. 38

Mazzaella linearis.

Constantinea (after Constance, a phycologist) consists of three species. Each plant sports a cup-shaped peltate blade (the stipe is anchored to the middle of the blade, like the leaf of nasturtium). The blade ranges in diameter from 6 to 12 cm (2½–5"). Branching is uncommon but may occur. Plants are found in the lower intertidal region in wave-exposed areas and in the subtidal region in a range of wave-exposed shores from Alaska to Mexico.

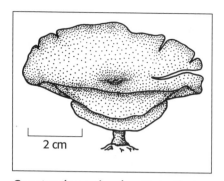

2 cm

Constantinea simplex. Note tiering of different year classes of blades.

A small nubbin is located at the top centre of the blade, directly above the stipe. Joe Powell (Sonoma State University, California) noted that the nubbin initiated growth in March, producing new stipe until September, at which time a new blade started to develop. This pattern of growth is a response to different photoperiods (length of day relative to night). The winter growth of the new blade is supported by storage compounds translocated (mobilized) from the old eroding blade. This occurs each year, leaving a scar on the stipe representing last year's blade. By counting scars, Powell determined the average age of *Constantinea* at Friday Harbor to be 6.85 years. This clever study was mostly concluded in 1964 under the direction of Professor Michael Neushul and published twenty-two years later. In his 1986 publication, Powell thanks Neushul "for his infinite patience in waiting for a publication on this work."

One species of *Constantinea* bears the lovely name *Constantinea rosa-marina*. Sandra Lindstrom advanced our understanding of this genus and related forms during her doctoral studies at the University of BC (1985).

Branched in one plane, flattened

In some species this feature is easy to recognize. However, in appearance other species seem to have features of both this flattened appearance and branching-in-the-round, which appears bushy. This morphological theme is found in green and brown algae also. It is interesting to speculate on what advantages this form may confer, compared to a blade. Enhanced nutrient supply, reduction of drag in wave-exposed areas and reduction of self-shading are some potential advantages.

Prionitis
Bleach weed
Photo 69, p. 39

Prionitis (Greek=jagged) represents up to nine species (three of which are common), all consisting of a mostly flattened branch system up to 50 cm (20") long. The side branches are frequently arranged as hairs along a feather. The plants are quite thick and tough. Their colour ranges from dark red to yellow, and the yellowish forms often reside in high tide pools or relatively quiet waters that may be low in nutrients. The red forms are found lower in the intertidal region in areas subjected to wave action that may provide more nutrients. These plants release a bleach-like smell when squished. This is thought to be a chemical defence against grazers. Representatives are found from Alaska to Mexico.

Opuntiella
Red opuntia

Opuntiella (Greek=diminutive cactus) is represented by one species, *O. californica*, which is very distinctive by virtue of its branching pattern. Roughly circular blades proliferate from similar blades, like a cartoon of the cactus *Opuntia*, reminiscent of Mickey Mouse caps. The entire plant may reach 25 cm (10") in length. These rich red plants are found in the lowest intertidal and subtidal regions in wave-exposed areas from Alaska to Mexico.

Osmundea
Red sea fern

Photo 70, p. 39

Osmundea (after Osmunder, the Saxon god of war) is a complex genus of approximately six species. Until recently this genus was known as *Laurencia*. *Osmundea spectabilis*, the most common species, has feather-like branching, which flows smoothly from branch to branch without abrupt joinings. The branch tips are rounded. The overall shape is reminiscent of a "pine tree" air freshener, or a fern leaf, particularly the fern *Osmunda*, also named after the god of war. The plants are usually less than 30 cm (12") long and may be

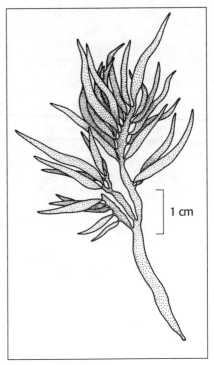

Prionitis lanceolata.

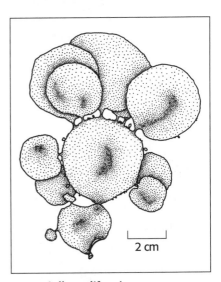

Opuntiella californica.

119

iridescent. They occur, usually in dense clusters, in the mid- to low intertidal regions in areas subjected to moderate to strong waves, from Alaska to Mexico.

Plocamium
Sea comb

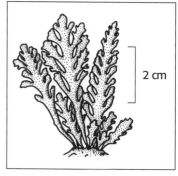

Osmundea spectabilis.

Plocamium (Greek=curled) is represented by three species whose curved profuse branches are restricted to one plane. The branching is distinctive: branchlets all arise from one side of their supporting branch, like teeth on a comb. The plants are clearly red and usually less than 20 cm (8") long. Representatives are found growing in the low intertidal region from Alaska to Mexico. Close inspection may reveal a small, tightly branched white or pale red algal parasite, *Plocamiocolax pulvinata* (see p. 102).

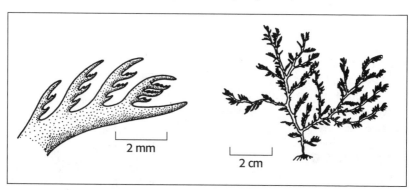

Plocamium cartilagineum and detail (left) showing branching on one side only.

Microcladia
Sea lace

Photo 71, p. 39

Microcladia (Greek=small branch) is a complex of three variously branched species, all having terminal branches pinched in like crab claws. Two basic branching patterns occur: branchlets arising from one side of the supporting branch (*M. borealis*) and branchlets staggered along both sides of the supporting branch (*M. coulteri*). Plants may reach 40 cm (16") in length but are usually less than 20 cm (8") long. Forms are often epiphytic on (attached to) coarse red and brown

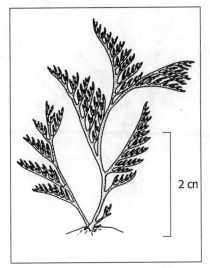 

Left: *Microcladia borealis*, showing branching on one side only.
Right: *Microcladia coulteri*, showing branching on two sides.

seaweeds in the mid- to low intertidal region from Alaska to Mexico. *Chondracanthus, Mazzaella* and *Prionitis* are popular hosts of *Microcladia*. There are other similarly branched red seaweeds but they are not as common.

Microcladia is a favourite for decorating note cards. See p. 22 for more on the art of seaweed pressing.

Ahnfeltiopsis
Flattened Ahnfelt's seaweed

Ahnfeltiopsis, a genus known as *Gymnogongrus* until recently, consists of three local species. Branching is a repeated forking of flattened, quite stiff elements, producing a fan-shaped system. Several such systems may arise from a red-crust base, resulting in a clump up to 20 cm (8") tall. Frequently, "naked swellings" appear on the branches. These house reproductive structures. *Ahnfeltiopsis* is distributed in the low intertidal and subtidal regions from Alaska to Mexico.

James Markham and Peter Newroth (earlier at the University of BC) explored the extraordinary ecology of the most common species, *A. linearis*, which inhabits the mid- and low intertidal regions of wave-swept beaches that are seasonally buried under 1–3 m (3'–10') of sand. On some pocket beaches, sand moves onshore in the spring and summer when wave action is gentle, and offshore in the fall and winter when wave action is strong. The offshore-moved sand is retained in an offshore sandbar.

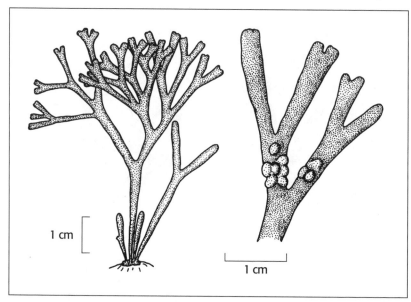

Ahnfeltiopsis linearis and detail (right) showing characteristic bumps.

Ahnfeltia
Bushy Ahnfelt's seaweed

Ahnfeltia (after Nils Otto Ahnfelt, a Swedish botanist who specialized in mosses) consists of two species of wiry, cylindrical and repeatedly forked branches that arise from rhizomes (prostrate branches). Their near-cylindrical branches may appear more bush-like than fan-shaped.

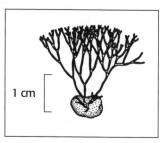

Ahnfeltia fastigiata.

Plants may reach 40 cm (16") in length but usually are less than 20 cm (8") long. These species are frequently found with *Ahnfeltiopsis* in sand-impacted areas, and form similar clumps. However, their cylindrical branches and rhizome attachment system distinguish these two genera. *Ahnfeltia* has a texture approaching a plastic scrubbing pad (but in good conservation practice should not be used for scrubbing). *Ahnfeltia* is distributed in the mid- to low intertidal regions from Alaska to Mexico.

Bushy branched

This common plant form results from branches arising in whorls, alternating pairs or any other configuration that deviates from a flat, two-dimensional system. The various branching systems that give rise to a bushy appearance can be seen among such plants as conifers, fruit trees and potted plants. The flattened branching system is rare among flowering plants.

Gelidium
Gel weed

Photo 72, p. 39

Gelidium purpurascens.

Gelidium (Greek=jelly) is represented locally by up to five species. Their terminal branches are usually flattened with branchlets arising like hairs on a feather. Lower down, the branches tend to be arranged around a central axis, giving the plant a bushy appearance. Plants are usually less than 40 cm (16") long and may be variously coloured from yellowish to deep red. Plants are usually restricted to the lower intertidal region and deeper, ranging from Alaska to Mexico.

Due to the quality of its agar, *Gelidium* is one of the world's five most economically important seaweed species (see Utilization and Cultivation of Seaweeds, p. 142). Our understanding of local species of *Gelidium* and its close relatives is the result of Dawn Renfrew's doctoral studies (University of BC, 1988).

Odonthalia
Sea brush

Photos 73, 74, p. 39

Odonthalia (Greek=toothed branch) is represented by up to five species, most of which are freely branched, having a distinctive bushy appearance. Some species may be partially flattened. The branch tips are sharp and often tightly clustered. Their colour ranges from brown to dark red to almost black. In tide pools and the mid- and lower intertidal regions, these plants take on a bewildering range of morphologies. However, the clustering of sharp-tipped branches seems consistent for the group. Mature plants may be small, from a few cm long to 50 cm (20") long. Representatives are found from Alaska to central California.

When observed at low tide on moderately wave-exposed shores, *Odonthalia* forms dense mats, which when parted are seen to cover an array of less conspicuous plant and animal life. The matted plants also house many zooplankton (floating animals). These small animals may

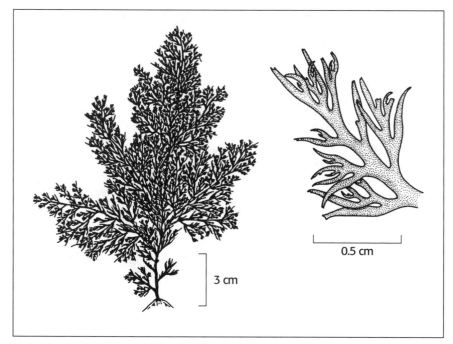

Odonthalia floccosa and detail (right) showing pinched terminal branches.

use the seaweed mat for protection from drying during times of low tide or they may live in the mat, acquiring their food from the microscopic plants living attached to *Odonthalia*. Tomas Probyn and Tony Chapman, Dalhousie University researchers, described a potentially mutualistic association between zooplankton and a bushy brown seaweed whereby the seaweed-inhabiting zooplankton nurture the plants with their excrement when nutrients are seasonally poor, thereby enhancing their home.

Neorhodomela
Black larch
Photo 75, p. 40

Neorhodomela (Greek=new red and black) is represented by three species. Their major branches are cylindrical and loosely branched. These give rise to numerous branchlets of similar length, with blunt tips, that are arranged in spirals. The plants appear brownish black and may reach 30 cm (12") in length. The blunt tips and arrangement of these branchlets reminded earlier phycologists of the conifer larch, thus the name *N. larix* for the most common species. These plants form dense, coarse mats, ranging from near-black to dark red, in the mid- and lower intertidal regions from Alaska to Mexico.

Neorhodomela often has a dusty appearance. This "dust" is diatoms, minute golden-brown plants (phytoplankton) that are normally found drifting in the ocean. These plants adhere to the seaweed. Rub a bit of *Neorhodomela* between your fingers and observe closely the particles that remain on your fingers. You may be able to make out small cylinders, rectangles and trapezoids. These are some common diatom cell shapes.

2 cm

Endocladia
Sea moss

Photo 76, p. 40

Endocladia (Greek=inside branch) is represented by one species, *E. muricata*, which is found in the mid- to high intertidal region from Alaska to Mexico. The dark-coloured plants are usually less than 10 cm (4") high. They stand out from the rock like little abrasive bushes. This abrasive appearance is the result of minute cone-shaped spines covering the irregular branches.

Neorhodomela larix.

Endocladia lives higher in the intertidal region than most persistent fleshy red seaweeds. Its ability to survive in an environment predominantly terrestrial (it spends more time out of the water than submerged) is attributed to its bushy nature, which enables it to retain moisture-laden air, thereby reducing overheating and desiccation. True terrestrial plants protect themselves by covering their leaf surfaces with waxy layers and stems with bark that hold in moisture.

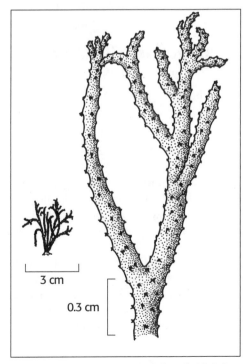

3 cm

0.3 cm

Gastroclonium
Sea belly

Photo 77, p. 40

Gastroclonium (Greek=belly branch) is represented by one species, *G. subarticulatum*, which occupies the lower intertidal

Endocladia muricata and detail (right) showing spiny surface.

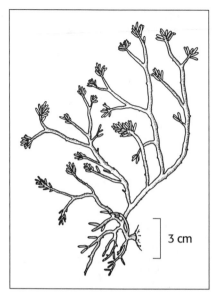

Gastroclonium subarticulatum.

region from Alaska to Mexico. Largish (up to 30 cm/12" long) plants are irregularly branched with the hollow terminal branchlets pinched into little plump sausage-like chains. Often the plants have reddish bases and yellowish tops.

Throughout much of its range, *Gastroclonium* is a major component of the low intertidal red seaweed turf. This assemblage of short, densely packed red seaweeds successfully defends its territory from kelp and other large brown seaweed competitors by persistently holding its space and invading new space through vegetative growth (like crab grass). The green and brown seaweeds are more dependent on spores that require exposed rock to invade new territory. In the past, the Japanese blasted intertidal rock free of this red turf community to encourage the growth of economically important kelp.

Gracilaria/Gracilariopsis
Red spaghetti
Photo 78, p. 40

Gracilaria/Gracilariopsis (Latin=slender) is a taxonomically difficult genus complex with up to four local species. This group seems to be distinguished by having no distinguishing features. Its sparse branches, which arise from a disc-shaped holdfast, are slender (1–2 mm/^{1}/32–^{1}/16" thick), and generally of uniform thickness throughout the entire plant, which may reach 1 m (3') in length. The colour ranges from brown to yellow to deep red. The plants are found in the low intertidal region attached to rock, and some are partially buried in sand. Others are found tangled with other plants on muddy shores. Their range is from BC to Mexico.

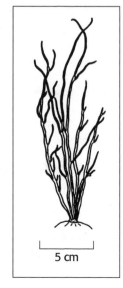

Gracilariopsis lemaneiformis.

Seaweed Ecology

The field of seaweed ecology is vast. The facets most relevant to our geographic region are discussed here; other aspects are presented in the descriptions and notes on the various species. For further information on seaweed ecology, see Annotated Selected Readings, p. 170.

Seaweed Distributions

Seaweeds exist in a great variety of oceanic conditions. They require sufficient light to support their activities and stable substrate to secure their attachment. Otherwise, seaweeds can thrive under tropical and Arctic sea water temperatures (-2 to 25°C/28 to 77°F), open ocean and estuarine salinities (3.5 to 0.3 percent salt content), severe and prolonged drying in the high intertidal region, and the stressful forces of surf waves. This is not to say all seaweeds can withstand all conditions. Some seaweeds only grow where waves are strongest, some may tolerate high temperatures, and others live under near-freshwater conditions. A few seaweeds do tolerate a wide range of environmental conditions and are found universally distributed, for example the green-bladed seaweed *Ulva* (see p. 50). These are referred to as cosmopolitan species.

Horizontal Distributions

On the large scale (equator to pole), sea water temperatures are thought to dictate patterns of seaweed distribution. Many of the seaweeds discussed in this book are distributed from Baja California, Mexico, through southeast Alaska. However, Point Conception, just south of Santa Barbara, represents a major break in west coast seaweed flora (see Biographic regions, p. 128). South of this point, there are many warmwater tolerant species that are not found north of Santa Barbara. Steve Murray (California State University, Fullerton) and colleagues have analyzed the seaweed flora in the vicinity of Point Conception. They noted that thirty species from farther north had their southern extreme near Point Conception and ninety species from farther south had their northern extreme near Point Conception. Between Santa Barbara and southeast Alaska the seaweed flora is relatively uniform, with most species distributed throughout. However, some species are restricted to either the northern or southern extreme of this coastline. Biogeographers refer to the coast from Point Conception through southeast Alaska as the Cold Temperate Region, south of Point Conception as the Warm Temperate Region and west of southeast Alaska as the Arctic

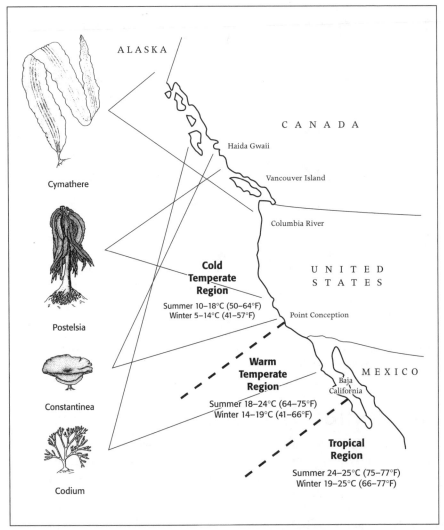

Biogeographic regions of the Pacific coast and their defining seawater temperatures. This guide covers the Cold Temperate Region.

Region. The general sea water temperatures defining the Cold Temperate Region are 5–14°C/41–57°F in winter and 10–18°C/50–64°F in summer (see Biogeographic regions, above).

We have one of the richest cold-water seaweed floras (as measured by numbers of species) in the world, which makes studying seaweeds here even more exciting. The much greater number of species in the North Pacific than in the North Atlantic is thought to reflect the greater age of the Pacific Ocean (more time to genetically diversify) and the greater negative impact of glaciation on the Atlantic flora. As time goes by and as we introduce Pacific species to the Atlantic (accidentally or on purpose), these differences will disappear (see *Sargassum*, p. 74, and *Codium*, p. 53).

Salinity, wave exposure and substrate type largely dictate local distribution patterns of seaweeds. Most seaweeds require stable substrate to exist. Sand, mud and small rocks that are tumbled by waves will not support attached seaweeds. A few seaweeds (see *Lola*, p. 45, and *Gracilaria*, p. 126) can exist unattached, but they are usually tangled with other seaweeds. Most seaweeds can tolerate a wide range of higher salinities (1.5–3.5 percent salt), but only a few can tolerate lower salinities. The greatest range in salinity is found in the high intertidal region and above, in splash pools—pools that are seldom flushed by the sea but receive some sea water via spray. These pools may bear crystalline salt in the summer as a result of evaporation and contain only rain-delivered fresh water in the winter, extreme conditions that are tolerated by a few green algae (see *Enteromorpha*, p. 52, and *Tetraselmis*, p. 55). Waves strongly influence the composition of local seaweed floras. Many species cannot tolerate the stresses associated with waves and will exist in more wave-sheltered waters "around the corner" or in the calmer, deeper subtidal waters. Some species exist only where the waves are strongest. They have adapted their reproduction and morphology to survive these stressful conditions (see *Postelsia*, p. 91, and *Costaria*, p. 88). At any one spot, where conditions of temperature, water motion, salinity and substrate seem favourable, the expected species may be absent. This local extinction may be the result of herbivory, poor reproduction or some recent stressful event such as an oil spill or high temperatures associated with El Niño. Or an "absent" seaweed may actually be present, but in another form (heteromorphic life cycle), or its presence may be abbreviated (an ephemeral or short-lived species). Conversely, a seaweed may be encountered in an unexpected situation. Some recent event may have allowed the seaweed to become established, such as La Niña, the irregular introduction of cold water; or there may be a local influence not present on adjoining beaches, such as up-welling—the introduction of deep, cold, nutrient-rich water to the shore. These anomalous distributions should alert you, the observer of seaweeds, to keep your mind open to the unexpected (see *Prasiola*, p. 49).

Vertical Distributions

The beach experiences a sequence of ebbings and floodings that is largely a function of the tides. Tide tables, in a book or the local paper, will provide you with predictions for high and low tides at any particular time and place. These predictions are based on astronomical information (relative positions of Earth, sun and moon) and the impact of local topography on tidal movements. For example, the

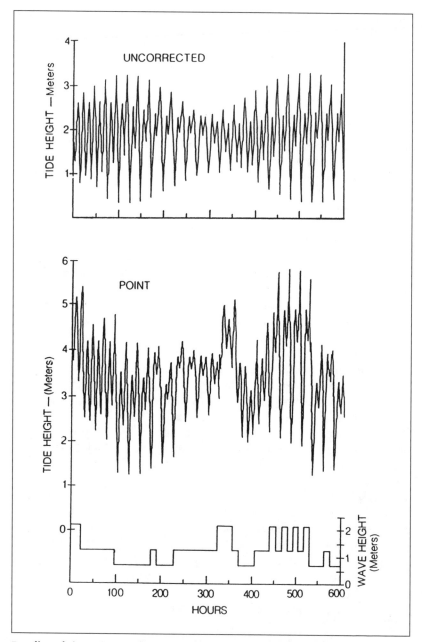

Predicted (uncorrected) and actual tidal heights for a wave-exposed point. Note the influences of wave height on the tidal pattern.

same astronomical influences are exerted on the west and east sides of Vancouver Island. However, there is almost a six-hour difference between tidal events in these two places, so if you are quick, you can collect seaweed on the same low tide on both coasts, even when it requires a three-hour drive and a two-hour ferry ride. The delay is caused by the constricted topography through which the tidal currents must flow. Tidal heights vary from 2.7 m (9') (San Francisco) to 7.4 m (24') (Prince Rupert, BC) over the area covered by this book. Superimposed on tidal forces and topography are the influences of weather conditions. Shifts in barometric pressure and winds may cause the tides to deviate substantially from the predicted values. The charts below illustrate the dramatic impact waves have on the predicted beautiful harmonic tide patterns for a wave-exposed beach. From these charts it is clear that the frequency and duration of any tidal exposure is not predictable by you or the seaweeds.

Most factors limiting the upper vertical limits of intertidal seaweeds have to do with meteorological conditions experienced when the tide is out. In the summer, exposed seaweeds may be subjected to desiccation and harmful unfiltered sunlight, including UV. In the winter, freezing—which may dry the seaweed as well as disrupt cellular structures—and exposure to fresh water (rain) may stress the seaweeds. The higher up on the beach the seaweed occurs, the longer the exposure to air, and the more severe the stress. Seaweeds positioned above mean sea level experience more exposure to meteorological conditions than to oceanographic conditions. As well as weather, grazing and biological competition for space and resources may limit the upward distribution of seaweeds.

As for the lower limits of species restricted to the intertidal region, some species may require exposure to meteorological conditions. Perhaps some drying is necessary to release spores or gametes. Other seaweeds may be limited in their downward extension by grazing or biological competition.

The upper limits of many predominantly subtidal seaweeds extend into the lower intertidal zone. When they are found there, exposure to meteorological conditions is probably infrequent and of short duration. They may be prevented from living higher in the intertidal region by conditions that are too dry or bright, or by grazing, or by competition for space and resources. The lower limits of subtidal seaweeds may be determined by biological interactions or inadequate light energy to support these photosynthetic organisms. Generally, the lower limits of subtidal seaweeds are much deeper in California than in Alaska, because less light reaches the greater depths in the north, where the water is less transparent than in the south. Northern waters are less transparent because they produce more phytoplankton, which intercept the light. When a seaweed's lower limits are determined by light, the seaweed is said to have reached its compensation depth, that depth where energy captured by photosynthesis balances the plant's energy requirements. That depth can place the seaweed in extreme environments. Mark Littler and colleagues recorded a crustose coralline growing at 268 m (884') deep off the Bahamas—the deepest depth confirmed for a seaweed. At this depth the algal crust received 0.001 percent of the surface light (about as close to dark as light gets). At a shallower depth (157 m/518') they recorded an unknown green seaweed that they subsequently named *Johnson-sea-linkia profunda* after the submersible used in this exploration and the benefactor Johnson (makers of baby powder).

Seaweed Communities

Seaweed species do not live in isolation, but rather in communities. The plant and animal composition of these communities is determined by interaction among members of the community, the physical setting of the community, and access to the community by outsiders. To identify and understand the multidimensional influences dictating community structure is a Herculean task, one we have only begun. Some areas of research have started to unravel the biological mysteries of seaweed communities. These relate to the physical structure of the seaweeds present, biological conditions such as competition and grazing, and the flow of energy among the community inhabitants.

Scientists have also studied succession, the progression of community development starting with lifeless substrate and concluding with a climax assemblage of species. Mark Littler, Diane Littler and Steve Murray have studied the biology of intertidal seaweed succession in California. The pioneering seaweeds, those that first colonize freshly exposed rock, are represented by *Ulva* (see p. 50), intermediate colonizers by *Egregia* and *Gelidium* (pp. 85, 123) and climax species by *Corallina* (p. 98). This succession represents a change of form from simple to complex. Generally, the earlier successional stages have higher production rates, and lower tolerance to grazing pressure and wave-induced drag forces than do the more advanced successional stages. The driving force for such patterns of succession are biological interaction (such as shading and grazing) and environmental factors (such as wave action), including the time of year when succession is initiated. One could assume the early successional species along a long stretch of beach that had just been devastated by a storm would be similar. However, after a period of time, differences in species composition would occur, reflecting micro-environmental differences along the beach and differences brought on by the developing community.

At first glance, the southern California giant kelp (*Macrocystis*) forests, which may run parallel to the beach for several kilometres, appear as homogeneous stands of this species. Closer scrutiny reveals considerable heterogeneity, in the form of fairly distinct patches within the forests. Different patches are dominated by different seaweeds, adapted to particular environmental conditions. Paul Dayton, Mia Tegner and associates (Scripps Institution of Oceanography, California) have considered the seaweed forest species composition within and between these patches. Some seaweeds are capable of quickly colonizing freshly exposed ocean bottom, as when the forest has been disrupted by storms or intense grazing. These seaweeds are called opportunistic, having high reproductive capacity, fast growth and usually short life spans (e.g. *Ulva, Desmarestia, Nereocystis*: see pp. 50, 72, 90). Some kelp, particularly *Macrocystis* (p. 93), are best suited to exploiting light and nutrients within the kelp forest. These plants form canopies high up in the water column that effectively capture light and nutrients. Still other kelp (e.g. *Eisenia, Dictyoneurum*, pp. 86, 89) are adapted to withstanding the severe physical drag associated with wave action. Finally, some seaweeds (e.g. coralline algae, p 97) are immune to grazing pressures. Thus there are sets of species capable of dominating a variety of micro-environmental habitats within the kelp forest. These environmental differences include changes brought on by the existing community, such as reduced wave stress and lowered nutrient levels in the interior of the forests.

Waves also play a role in determining the composition of seaweed species inhabiting the beach. For example, *Postelsia, Alaria nana* and *Mazzaella cornucopiae* (pp. 91, 82, 116) are found only on beaches exposed to strong wave forces, and *Zostera, Laminaria saccharina* and *Alaria tenuifolia* (pp. 42, 78, 82) are found only in wave-sheltered situations. Most seaweed species are found between these two extremes of wave exposure. Some of them modify their morphology to accommodate the prevalent wave environment. Usually they become narrower and thicker in wave-exposed situations, relative to their morphology in wave-sheltered conditions (see *Costaria*, photo 38, p. 32). This modification is termed phenotypic plasticity, which implies the morphological change is environmentally induced and not an expression of the plant's genetics. However, the ability of a plant to express phenotypic plasticity is apparently an inherited trait.

Mimi Koehl (University of California, Berkeley) identified the mechanism that allows *Nereocystis* (p. 90) to survive the drag forces associated with waves and currents. As a wave flows past the kelp, tugging on its float and blades, the stipe absorbs the stressful energy by stretching— up to 30 percent of its length. The result of this wimpish behaviour (Koehl's phrase) is that by going with the flow, the bull kelp can withstand the same forces as bone and wood withstand by being rigid. Koehl is also a skillful choreographer. I once had the pleasure of participating in her creation of Bull Kelp and the Nero Sisters (done in Motown style), performed at an annual Invertebrate Ball at Friday Harbor. (The event was judged by Rae Hopkins, who later became my wife.)

Grazing exerts strong pressure on seaweed assemblages, and seaweeds have several strategies to cope with this menace. They may be opportunistic, quickly occupying available space, developing and reproducing before the animal onslaught (e.g. *Ulva, Desmarestia*). Their tissues may be too tough for animals to assault (e.g. woody kelp stipes, coralline algae, some soft crust forms). They may support their soft tissues high in the water column, out of reach of bottom-dwelling herbivores (e.g. *Macrocystis, Pterygophora*). They may produce a chemical defence to deter herbivory (e.g. *Desmarestia, Prionitis, Fucus*, many kelp). Without question these strategies are effective, but they may or may not have evolved to protect the plants from herbivore damage. For example, some chemicals that deter herbivory also protect the plant from UV damage. It is very likely seaweeds had to deal with harsh UV conditions before they encountered sophisticated herbivory. Similarly, scientists believe insect wings may have evolved as solar collectors, to warm the animal, and later became adapted to flight.

On the west coast of North America, the impact of some human activities on seaweed community structure can be traced. Studies by David Duggins (Friday Harbor Laboratories, Washington), James Estes (University of California, Santa Cruz), Jane Watson (Malaspina University College, BC), and others have painted the following picture. In the beginning there were hunters (humans), predators (sea otters), herbivores (sea urchins, sea cows) and seaweed. Sea cows were hunted to extinction in the eighteenth century. We know little about their impact on seaweed, but we may assume they required large amounts of food to support their bulk. By the end of the nineteenth century, sea otter had been hunted to extinction along much of the Pacific coast. Without the sea otter to prey on them, sea urchins multiplied, increasing the grazing pressure on the seaweed community.

Generally, sea urchins graze at subtidal depths that are below the major influence of surface waves. This focusses grazing stress on the lower limits of subtidal seaweeds. Thus, where the lower limits of a particular species might have been determined by its light requirements, its lower limits are now determined by the upper limits of sea urchin distribution. In Barkley Sound, on the west coast of Vancouver Island, *Macrocystis* (p. 93) is generally restricted to about 2 m (6$\frac{1}{2}$') below low tide level, the upper limits of the majority of the sea urchins. When sea urchins were excluded from these areas in experiments conducted by Dan Pace (Simon Fraser University, BC), *Macrocystis* grew to depths of 8–10 m (26–33') below low tide, and the kelp bed grew much larger.

In the mid-twentieth century, sea otters were reintroduced to parts of BC. Jane Watson monitored the impact of these reintroductions and found that generally sea urchin populations declined and seaweed grew to lower depths. Seaweed flora shifted from crustose coralline algae and opportunistic species (pre-sea otter) to canopy kelp and stiff-stiped

A kelp bed food web, showing the relationship of various biological components.

kelp (post-sea otter). However, the pattern of community response to sea otter reintroduction was not always predictable.

The questions of predictability and generalization in ecology are important. At what point may an ecological event be considered predictable? How often and at how many different locations must you observe an event to establish its predictability? Statistically, we may establish the probability of an event in marine life occurring, much as a pollster will predict the outcome of an election. However, statistics require large samples—that is, a large number of observations—which are hard to come by when complex ecological situations are being studied. So then how are we to proceed? Questions such as community response to the reintroduction of the sea otter are not just of academic interest; they are crucial to managing near-shore fisheries such as the abalone and sea urchin fisheries. In many instances, introduced species have not behaved in the expected manner, resulting in ecological disruption. The sea otter is not a battery-driven toy, performing the same activity over and over and over. It is an intelligent animal that may learn to exploit new food sources and pass this "knowledge" on to its pups, so that the reintroduced sea otter may bring with it new hunting skills and food preferences. These and similar questions of predictability and generalization are under intense debate by Robert Paine, Mike Foster (one of Neushul's students), and others.

It is critical that we understand those processes that dictate the composition and stability of marine (and terrestrial) communities. Human beings leave an impact on practically every aspect of Earth's environment, usually a negative one. Only tides and other gravity-related matters and seismic and other geomorphology-related matters have escaped our influence. The response of an affected community reflects, in part, the roles of the various species within the community. Three general roles have been suggested. First, the "key stone species" hypothesis states that an individual species' activities have a profound impact on the overall community structure. For example, the sea otter may be considered a key stone species. Its presence may reduce sea urchin grazing, allowing seaweeds to increase. The second theory is that several species are essential to the integrity of the community, although we do not know how many. In this, the "airplane rivet hypothesis," the question is how many rivets an airplane can lose before crashing (but let us not dwell on this one!). The third theory is that numerous species play the same crucial role, thus insulating the community from harm. This is known as the "redundancy hypothesis."

Seaweed Productivity

A look at the activities of seaweed communities can suggest that seaweeds are at war with each other, with marine animals and with their physical and chemical environment. In reality, they play a crucial supporting role. They are the major primary producers in our shallow coastal waters. The light energy they chemically capture, via photosynthesis, is distributed to marine animals directly in the form of intact seaweed tissue (they are grazed upon), and indirectly as fragmented plant tissue (detritus) and as dissolved components of the seaweed. Ecologists usually evaluate the primary productivity of a plant in terms of how much carbon it captures and incorporates into organic compounds per square metre (sq yard) of ocean surface per year (this is written as $gC\ m^{-2}\ year^{-1}$). Seaweed productivity in our region compares well with the high productivity of cultivated alfalfa in Puerto Rico:

Some Primary Productivity Values

	$gC\ m^{-2}\ year^{-1}$
Nova Scotia kelp beds	1,750
British Columbian *Macrocystis*	1,300
Hawaiian crustose corallines	560
Alaskan *Zostera*	1,450
Coastal phytoplankton	200
Puerto Rico alfalfa	2,100

Source: K.H. Mann (1973), *Science* 182: 945.

Whereas there is little question kelp forests and eelgrass beds are sites of high production along our shores, there are questions as to the fate of their energy-rich compounds and their significance in supporting near-shore communities. Consider that seaweed communities comprise a vegetative band usually less than 100 m (330 ft) wide, running along our shores. And consider that that band is a very small portion of a much larger oceanic area that supports a microscopic population of primary producers (phytoplankton). Noting the productivity values given above, it is easy to see that the vast expanse of open ocean greatly outproduces the coastal fringe. Duggins and colleagues (University of Washington, University of California, Santa Cruz) were able to establish the source of carbon that supported a variety of near-shore animals by examining their tissues. Kelp and phytoplankton, the two major primary producers in the study area (Aleutian Islands), incorporate different isotopes of carbon in different ratios in their plant tissues. These same distinct carbon ratios were found in the herbivore tissues and, in turn, in the tissues of their predators.

Distribution of Kelp-Originating Carbon Among Other Organisms

Organism	Feeding mode	% Carbon of kelp origin
Mussel	filter feeder	38
Barnacle	filter feeder	85
Crab	detritus feeder	38
Rock greenling	predator	68
Sea star	predator	51
Cormorant	predator	48

Source: D.O. Duggins, C.A. Simenstad and J.A. Estes (1989),
Science 245: 170–173.

The average kelp carbon content of the various animals studied ranged from 15 percent (a subtidal sea anemone) to 95 percent (a subtidal zooplankton species). This carbon was found in all animals tested, including cormorants (48 percent) and predatory fish (68 percent). From this study it seems reasonable to conclude that near-shore seaweed primary production, in combination with the greater phytoplankton production, is an important energy source for shallow coastal water animal communities. What role does kelp primary production play in supporting offshore animals and animals that migrate through shallow coastal waters (salmon, herring)? These and related questions are being pursued.

Conservation Biology of Seaweeds

Conservation biology is the scientific response to biodiversity crises—crises brought on by the impact of human activities on our natural environment. The discipline, a product of the 1980s, is a synthetic field comprising established areas of ecology, population genetics, economics, sociology, philosophy and others. A phrase often associated with conservation biology is "biological diversity," which has been defined by the US Office of Technology Assessment as "the variety and variability among living organisms and the ecological complexes in which they occur."

When the popular media present information on biodiversity, they often focus on crises in the rain forests, the discovery of new exotic life forms associated with submarine thermal vents, the loss of water-purifying wetlands and other sensational situations. Our shallow coastal waters may not be so media-worthy, but they are beleaguered as well. Human foraging for intertidal life and innocent trampling along the beach have significant damaging effects on this fragile ecosystem. Sewage outfalls, thermal electric generating plants, large oil spills and small ones (e.g. small amounts of oil released from jet skis), destruction of the ozone layer (allowing increased amounts of harmful UV to reach the earth's surface), and global warming all stress the inhabitants of our shores. None of these stresses is natural.

Our response to the seaweed beach crisis has been slow but methodical. Early and ongoing efforts have focussed on inventorying beach biota—what species are where and how abundant they are. Coupled with the basics of inventory are studies monitoring change in beach flora over time, with the idea of establishing a correlation between biological change and introduced stresses. Finally, a few studies have tried to measure the impact of human-induced stresses on the plants of the beach.

First Nations people compiled the first seaweed inventories. Unfortunately, much of this knowledge has been lost and only fragments remain (see Annotated Selected Readings, page 170). The post-contact scientific inventory phase has developed more or less haphazardly. On the north Pacific coast, this phase effectively began with the observations of A. Postels and F.J. Ruprecht, two mercenary marine botanists in the employ of the czar of Russia, in the early 1800s. Following their early visits, our knowledge of flora in the area grew through individual efforts, usually geographically confined. William A. Setchell and N.L. Gardner, working out of the University of California, Berkeley, greatly expanded our understanding of seaweed demographics along much of our coasts in the early 1900s.

After that, the next major work was undertaken in a series of cruises over four years in the 1960s. The project, which employed Canadian Navy auxiliary vessels, was sponsored by Robert Scagel. Each cruise lasted about six weeks. Two cruises systematically inventoried the seaweed flora from southern BC to Alaska and the other two inventoried the flora from southeast Alaska to the northeast tip of the Aleutian Islands. I was privileged to be a part of three of these cruises. The thrill of waking up every morning, anchored off a yet-to-be-sampled beach and anticipating new discoveries, was beyond description. The thousands of seaweed specimens collected on these cruises are housed in the University of BC herbarium, Vancouver, where they are available to students of biodiversity.

Serious studies that track changes in seaweed species diversity and abundance are now underway. Megan Dethier initiated a series of intertidal transects in Washington state in 1988, designed to monitor changes in the distribution and abundance of seaweeds and invertebrates. These transects, which focussed on areas subjected to little human-induced stress, have been expanded and continue to be monitored by the US Parks Service. Some of these transects may be taken over by PISCO (Partnership for Interdisciplinary Studies of Coastal Oceans), which is studying biological diversity and ecosystem function along the west coasts of Washington, Oregon and California. This partnership is financed by a private grant from the David and Lucile Packard Foundation for $17.7 million US.

Megan Dethier's early interpretation of data from these transect studies is that there is considerable year-to-year variation in species distribution and that the change seen usually does not conform to any particular pattern. She suggests the value of these studies is to come to grips with the range of possible variation encountered in unstressed situations, thereby defining an "envelope of normalcy." This envelope may then be used to assess the impact of a defined stress.

A few studies have explored the impact of well-defined environmental stress by monitoring the response of the seaweeds affected. In central California, John Steinbeck, David Schiel and Michael Foster have monitored seaweed distributions and abundances for the years 1976–1995. Their purpose was to observe changes in seaweed flora before and after the opening of an electric power plant that discharged heated sea water into the study site. The first ten years of the study took place before the plant opened, providing much-needed background. The remaining years of investigation followed changes in the seaweed flora after the heated water began to be discharged. The study documented a drastic shift in the species composition of the affected seaweed flora—more than 100 species changed distributions. The researchers had expected that after the plant opened, the seaweed flora would shift to resemble the species mix found farther south, where water temperatures were similar to that of the water discharged by the plant. This did not happen. The major kelp, including the large bull kelp (*Nereocystis*), disappeared, with the exception of the giant kelp (*Macrocystis*), which increased in abundance. This shift in kelp species could have a profound impact on the habitats available to fish and invertebrate life, resulting in a shift in other species composition.

On March 24, 1989, the oil tanker *Exxon Valdez* crashed, losing its cargo of 30 million litres of oil and wreaking havoc in Prince William Sound, Alaska. It was one of the worst human-induced environmental disasters in North America. Exxon was fined $250 million US for settlement of criminal charges and $900 million US for settlement of civil charges. Such settlements are based on evaluations of damages done, but how can the real damage be calculated? How much is a sea otter worth? How many limpets were killed and what is their value? What is seaweeds, any seaweed, worth?

The *Exxon Valdez* oil spill and its effects were the basis of an extensive environmental impact study. However, one crucial element was missing: there was no pre-impact baseline study with which to compare data. Numerous studies were undertaken to understand the impact of the disaster on seaweeds. They were led mostly by Michael Stekoll (University of Alaska), Sandra Lindstrom (University of BC) and Gayle Hansen (Hatfield Marine Science Center, Oregon). The first result of these studies was that it was generally not possible to distinguish between the impacts of oil spillage and the impacts of the subsequent cleaning up. (Exxon was forgiven $150 million US of settlement charges, mostly in recognition of its co-operation in the cleanup.) In any case, the impact on intertidal plants and animals was severe: some species required three to five years to become reestablished. Much valuable knowledge was acquired from studies on this spill, and monitoring of Prince William Sound continues.

Scientists have also studied the impact of people paying visits to rocky intertidal shores. Steve Murray and colleagues (California State University, Fullerton) monitored human behaviour and found that humans affected the beach more or less in proportion to the density of the local population size. There was little difference in the frequency of visits or intensity of collecting activity on beaches designated biological reserves and those not designated as reserves.

A major impact of human visits on rocky intertidal beaches is trampling. Deborah Bronsan (Oregon State University) and Lana Crumrine

(University of Oregon) established experimental plots subjected to trampling and no trampling treatments. The trampled plots were tread on 250 times per month for one year and then allowed to recover for another year. When the trampled plots were compared to their controls (the untrampled plots), it was observed that bladed seaweeds such as *Fucus* (p. 67) and *Mastocarpus* (erect phase) (p. 110) suffered significant declines in the trampled plots. On the other hand, little bushy plants such as *Gelidium* (p. 123) and *Endocladia* (p. 125) and algal crusts such as *Mastocarpus* (crust phase) (p. 101) benefited from trampling. This study nicely confirmed results of an earlier study conducted by the two Oregonians, who noted that coastal areas subjected to frequent human visits had a preponderance of crust and bushy seaweeds in contrast to areas not frequently visited. Further, when a trampled area was protected from human visits, the seaweed population shifted back to a balance between the bushy/crust species and bladed species.

All of these studies demonstrate the direct impact we have on our seaweed flora. At one extreme, the *Exxon Valdez* oil spill, essentially all the intertidal seaweeds over an extended geographic range were destroyed for several years; and at the other extreme, trampling caused a shift in the morphology of the dominant seaweed species. In all cases, the nature, breadth and duration of this impact are not known, particularly the cascading effects. For example, what happens when a fish is deprived of an egg-laying habitat because of human trampling? What is the impact of this deprivation on other species that prey on the fish? These cause-and-effect relationships are very difficult to analyze and predict. Often we may have more than one piece of the puzzle and not know it, or not understand how they fit together. Consider: could the deterioration of a coastal kelp forest and the loss of an offshore fin fish fishery be connected through cascading events? What events might arise from floristic shifts resulting from global warming, where the prevalent temperatures might shift from those normally encountered in our region (Cold Temperate Region) to those encountered farther south in the Warm Temperate Region (see figure on p. 128).

Utilization and Cultivation of Seaweeds

This section emphasizes contemporary seaweed products and their production on the Pacific coasts of North America. Seaweeds have a global impact and affect our lives every day. For example, plant gums extracted from seaweeds are used to keep the chalk in toothpaste suspended. Were it not for these gums the toothpaste would consist of perfumed water (perhaps with fluoride) and a lump of chalk. These same gums are used to size cloth, clear beer, form dental moulds and emulsify oils and fats in a variety of food, cosmetics, medical and industrial products.

Nursery farms for early stages of nori farming in Puget Sound, Washington.

In the past, seaweeds were considered specialty food products for a minority of North Americans. Stores catering to people of Asian descent sold various kelp species (under the names kombu, wakame—Japanese; hadai—Chinese) and a red alga *Porphyra* (nori—Japanese). On the east coast of North America, dulse was marketed locally as a chip. Health food stores carried some of these products and usually kelp pills. Using seaweeds as fertilizer was an informal exercise, usually restricted to gardeners living near the seashore. On the west coast, Native peoples harvested herring roe attached to seaweed and other substrate. This was used as a special feast food and as a trade item with inland Native groups.

In the last twenty years, there has been an explosion of new seaweed products for North Americans and a growing local acceptance of these and more traditional products. Also the technology to farm some species of seaweed for commercial purposes has been developed.

World Seaweed Production

The demand for seaweed products is growing as seaweeds are used in new ways and as Earth's population increases. Lindsey Zemke-White (Auckland, New Zealand) and Masao Ohno (Kochi, Japan) have provided "an end-of-century summary" of our seaweed production, based on 1994–95 data.

Seaweed group	Dry metric tons	Percent cultivated
Green seaweeds	5,998	75
Red seaweeds	1,042,507	23
Brown seaweeds	956,954	83
Total	2,005,459	52

Compared to a similar study conducted for the year 1984, this 1994–95 production represents a 119 percent increase in overall seaweed production.

Seaweed production for 1994–95 from Pacific shores, from the Aleutian Islands to the southern extreme of Mexico, was 25,370 dry metric tons or 1.3 percent of the world total. Less than 1 dry metric ton of this production was from cultivated plants.

Food

The single most important seaweed food is nori (*Porphyra*), which is used to wrap sushi. Initially, this seaweed food was used exclusively in Asia, particularly Japan. Nori has been popularized throughout the western world, probably starting from California, where it was incorporated into CalAsian cuisine. Today, the annual value of cultivated nori is approximately $2.8 billion US.

Porphyra farming is done in two phases. First, nets are seeded with spores from the filamentous phase of *Porphyra* that has been cultivated in seashells under controlled greenhouse conditions. These nets are then planted in the sea, where they produce the bladed stage of *Porphyra*, and then the nori is harvested. Tom Mumford (Washington State Department of Natural Resources) was instrumental in introducing this method of *Porphyra* cultivation to North America. John Merrill (Applied Algal Research and later Michigan State University) started the first nori farms on this coast, using a Japanese species. These farms appeared to be economically viable but local residents insisted that they be removed, citing visual pollution as the problem. Similar farming attempts in BC have failed, but apparently not for biological reasons. North American nori farming is now restricted to Maine. Robert Waaland (University of Washington) has explored the biology of local *Porphyra* species with the aim of facilitating their cultivation in the future.

Nori farming: harvesting mature nets in Puget Sound, Washington.

Edible kelp constitute the second most valuable sea vegetable. Farmed and wild harvested plants have an annual value of approximately $270 million US. *Laminaria*, both wild and cultivated, is an important sea vegetable in Asia. Aquaculture of *Laminaria* was initiated in Japan and then in China. The Chinese story is fantastic. The waters of the Yellow Sea are too warm for *Laminaria* survival, so the Chinese reared *Laminaria* in refrigerated pools, where they subjected the plants to various mutigens. Eventually they achieved a warm-water tolerant variety and initiated massive culture in the sea. Whole communes are dedicated to producing kelp "seed," to maintaining the farms and to harvesting and processing the plants. As a result, China is able to meet the iodine requirements of approximately one quarter of Earth's population from farmed kelp, with an annual farm production of 644,000 dry metric tons.

In addition to traditional Japanese kelp products and nori, a growing list of North American seaweed products is aimed at the North American consumer. About six companies, mostly small, service and compete for local markets. Growing demand for these products reflects the greater exposure of North Americans to sushi shops, western-oriented sea vegetable cookbooks and the spread of information on related health benefits. In the future there may be a demand for North American kelp in Japan, as its population expands and demand for seaweed products outstrips domestic supply.

As in *Porphyra* farming, the cultivation of kelp consists of two phases. The first phase involves the seeding of string under controlled

Kelp "seed" on string, ready to plant out.

145

laboratory conditions. The string is seeded with small (usually less than 4 mm/⅛" long) kelp blades, which have grown from artificially fertilized microscopic female plants. This string is then attached to ropes that are suspended in the sea. North American kelp cultivation was initiated in BC in the early 1980s and soon thereafter in Puget Sound, Washington. The technology driving this cultivation was introduced from Japan and modified to fit local conditions and recent advances in our understanding of kelp biology.

The basic farm unit employed in BC and Puget Sound consisted of an anchored 40x70 m (130x230') rope rectangle suspended 2 m (6.5') below the surface. This frame, which is divided lengthwise into two equal rectangles, supports sixty 20-m (66') long farm ropes. Production by this type of system ranged from 3 to 28 wet kg kelp per metre of rope (2–19 lbs per foot) for various species of *Laminaria*, or 3.6 to 33.6 metric tons (4–37 tons) per unit.

In 2001 there were two kelp farms operating in BC, one for kelp production for homeopathic pharmaceuticals and the other in support of a sea urchin feeding operation. In addition there were four test farms in support of the herring-roe-on-kelp fishery. The following species have been successfully cultivated in Puget Sound (John Merrill) and/or BC (Rae Hopkins and Louis Druehl, Canadian Kelp Resources Ltd.): *Laminaria saccharina, Laminaria groenlandica, Nereocystis luetkeana* and *Macrocystis integrifolia*. Attempts to farm *Alaria marginata* have been unsuccessful due to severe grazing by a small crustacean.

Four-month-old *Laminaria saccharina*, ready to harvest.

Pharmaceuticals

North Americans are becoming more aware of potential health benefits associated with kelp, a perception increasingly supported by research. French researchers reported that a substance extracted from *Ascophyllum nodosum* (a fucoid, related to our *Fucus* and *Sargassum*) "is a very potent antitumor agent in cancer therapy." Other researchers have demonstrated the potential efficacy of kelp-derived constituents in the treatment of high blood pressure, stroke and coagulation, and aspects of diabetes. Kelp products also have anti-inflammatory effects.

All of the above studies were conducted on rodents or human tissue in culture. Much work needs to be done to identify the active constituents and to test them extensively. Pills made from brown algal powders, which contain a healthy array of micronutrients, are currently the major seaweed pharmaceutical. It is clear that seaweeds, particularly brown algae, may prove to be sources of valuable pharmaceuticals.

Cosmetics

The use of seaweeds as cosmetic aids or sources of cosmetically active compounds is a growing area of interest. Brown algae (usually *Laminaria* or *Ascophyllum*) are commonly used in the manufacture of beauty aids. Seaweed and seaweed extracts are often used in thalassotherapy, a form of skin care involving sea water, which has a long tradition in European spas. In fact, it was used to cure just about everything that might ail you. Today seaweed is used in "mud baths" or "body wraps" because it helps exfoliate dead skin, rejuvenating the bather.

Seaweed and seaweed extracts are also found in a variety of skin creams and lotions, and shampoos. Some of these products are made locally, on a small scale. In North America the use of seaweeds is a small but essential part of the spa industry, but elsewhere, particularly France, the development of seaweed-based beauty aids is exploding.

Another seaweed product that has received much attention is the fat patch. This small patch, which contains an extract of *Fucus*, is used in losing weight. The claim is that the patch introduces iodine into the body, stimulating the thyroid gland and speeding up metabolism. The method has its supporters and detractors.

Fertilizers and Pesticides

Seaweeds, mainly kelp, harvested from the sea or the beach are variously treated (powdered, extracted) and sold as fertilizers to be sprayed on or added to the soil. These seaweed materials may promote plant health by providing growth regulators and pesticidal components, as well as supplying nutrients. When subjected to scientific scrutiny, such claims do not always hold up, but many claims are valid and users are convinced of the product's worth.

The increasing demand for organically grown fruits and vegetables has driven producers to seek new ways to fertilize, control pests and regulate plant production that meet the requirements of certified organic farming. Further, decades of farming have depleted important micronutrients from soils and other farmers are using seaweed-based fertilizers to restore their farms.

Sea Urchin and Abalone Feed

Sea urchin roe is a valuable fishery. In 1996 red sea urchins had a landed value in BC of $11 million Canadian ($7.7 million US). Much of the wild harvest in BC and elsewhere is stressed, and there is a strong move to cultivate animals. In BC two models are being pursued: 1) tank-rearing through the entire life cycle and 2) field ranching of natural populations. In tank rearing, the animals must be fed through their entire growth period, and in ranching, the animals (or rather their roe) are fattened for a few months before being harvested.

The animals considered for ranching in BC exist on algal barrens where natural food is very limited and the urchin population manages to support itself with poor quality roe. A study conducted by Canadian Kelp Resources Ltd. with the Huu-Ay-Aht First Nations indicated that a 1-acre (.4 ha) farm of *Laminaria saccharina* could feed a minimum of 44,000 urchins, of which 13,000 would be of harvestable size, for one month prior to the roe harvest. The estimated maximum value added to the harvested roe due to increased quality would be 145 percent or $16,200. In addition, there is the advantage of having strengthened 31,000 undersized urchins for future harvests. The study included test feeding of algal-barren sea urchins with farmed kelp, demonstrating the logistical feasibility of this approach.

Wild stocks of Pacific abalone have been seriously depleted. Fisheries authorities in BC have imposed a ten-year moratorium on wild harvest. The rearing of seaweeds to support tank culture or ranching of abalone presents an exciting opportunity for coastal residents.

In supporting lucrative sea urchin and abalone fisheries, kelp may be used as an ingredient in artificial diets, a manipulated farmed food or an enhanced wild food. From my perspective, kelp should always be considered the optimal food. Increasingly, today's consumers prefer animals fed naturally to those fed artificially. Currently there is one commercial kelp farm in BC supporting a commercial sea urchin rearing facility.

There have been two local studies into obtaining seaweeds using waste products from cultivated animals. Royan Petrell and colleagues (University of BC) have been investigating the feasibility of farming *Laminaria saccharina* downstream from floating salmon farms. They asked: Is it possible to enhance kelp growth with salmon-originating nitrogenous wastes while cleansing the overly nutrified seawater? Based on early results, the tentative answer to this question is that the kelp would show enhanced growth but there would be little lowering of the nitrogen in the sea water.

Aquaculture studies conducted by John-Eric Levin and William McNeil and most recently by Ford Evans and Chris Langdon (Hatfield Marine Science Center, Oregon), have demonstrated the feasibility of rearing *Palmaria* (the seaweed from which the seaweed chip dulse is derived) in salmon farm effluent. The dual advantages of this system are a cleansing of the effluent and the production of a cash crop (edible dulse and/or abalone feed). This study and that of Petrell's group are among the first serious studies of marine poly-culture on our coasts.

Herring-Roe-on-Kelp and other Seaweeds

Herring-roe-on-kelp (HROK) is a popular delicacy in Japan. It is produced in Alaska, BC and Washington. In BC, the HROK fishery is worth approximately $20 million Canadian ($14 million US) annually. This fishery involves bringing together quality herring and kelp (*Macrocystis*) in an enclosure, which is often located far from quality kelp, so kelp must be transported. Excessively handled kelp loses its ability to adhere to the herring roe, resulting in an inferior product. Farming may solve some problems by providing kelp in the vicinity of the herring pens. Also, through selection, superior kelp may be provided to this fishery.

Traditionally, herring roe is harvested from slaughtered fish. This is much like killing the golden-egg-laying goose, as herring may spawn many times. The roe-on-seaweed approach allows the herring to continue spawning through its natural life span (up to seven years). Other seaweeds may be used to expand this more sustainable approach to the roe fishery. *Gracilaria* is used as a source of agar and eaten as a sea vegetable in Japan (ogo nori) and Hawaii (limu manauea). There has been an attempt to farm *Gracilaria* in raceways in Hawaii to supply local markets. Isabella Abbott, while at the Hopkins Marine Station, Stanford University, explored the possibility of producing herring-roe-on-*Gracilaria* as a product for human consumption.

Seaweed Gums

Three major plant gums are derived from seaweeds: agar, carrageenan and algin. Agar and carrageenan are derived from red seaweed and algin is derived from brown seaweed. These gums, which are also referred to as phycocolloids and seaweed polymers, are associated with seaweed cell walls. They exist in a variety of subtle forms, which are distinguished by their different gelling properties. This range of properties gives the manufacturer a wide range of compounds to work with. The gums are used mainly as gels, thickeners and stabilizers, and their total annual value was estimated to be $2.6 billion US in 1994.

Gelidium is a major source of agar, which because of its jelly properties has many uses. For example, lower-grade agars are used to pack canned meats, middle grades are used in preparation of microbiological media, and high-quality agar is used in gels for manipulating DNA in various areas of biotechnology and criminology. This high-grade agar is very expensive ($1 US/gram). At least one Pacific coast company is attempting to produce top-quality agar through genetic selection and cultivation.

Gracilaria is another significant source of agar. A floating bag system has been designed by Geoff Lindsay and Rob Saunders (Bamfield, BC) to cultivate this agarophyte. The culture bags are suspended in the sea and supplied with nutrient-rich artificially upwelled sea water. This deep sea water is delivered via compressed air. The system was technically sound, but the economics of the day (about 1980) made it impractical.

Carrageenan may be derived from the local red seaweeds *Chondracanthus* and *Mazzaella*. This gum is particularly useful in the production of milk-based products, such as chocolate milk. Robert Waaland has investigated the feasibility of farming these seaweeds in Puget Sound, Washington.

The cultivation of red seaweeds for their plant gums is impractical in our local waters, mostly because of the lower costs of producing them elsewhere, particularly in warmer regions. Possible exceptions to this are the production of particularly novel and high-valued products such as biotechnical grade agar.

Algin is the only marine plant gum produced in our local waters in significant quantities. Kelco, a subsidiary of Merck, harvests up to 140,000 wet tons of *Macrocystis pyrifera* yearly in southern California. This wild harvest, which was initiated in 1911, does not appear to have harmed the kelp beds, although it must be pointed out that there has been no control kelp bed with which to compare this century-long harvesting experiment. However, infrequent sea urchin invasions and warm water intrusions (El Niño) have decimated local kelp patches. Wheeler North (California Institute of Technology) has been largely responsible for providing the scientific understanding necessary to restore these damaged beds.

Biogas from Kelp

Methane gas is produced by the anaerobic digestion of organic matter by a variety of specialized bacteria species. This gas may be burned directly for heat or used to power electrical generators. The digester employed can be very simple, essentially an insulated airtight container with an import for fuel and an export for trapping the gas. Various fuels can be used, and kelp is one of the highest producers of methane gas.

Methane Production Values

Fuel	m^3 of methane per kg of dry organic matter
Raw Kelp	0.28
Municipal solid sewage	0.10–0.24
Feedlot cattle waste	0.17–0.26
Dairy manure	0.21

Source: D.P. Chynoweth, S. Gosh and D.L. Klass (1981). "Anaerobic Digestion of Kelp," in *Biomass Conversion Processes for Energy and Fuels*, eds. S.S. Sofer & O.R. Zaborsky, NY: Plenum Press, pp. 315–38.

In the late 1970s, the oil producing and exporting countries (OPEC) drastically raised the price of oil, driving energy costs sky high. The US responded in part by exploring alternative energy sources, including biogas from kelp. Wheeler North (California Institute of Technology) and Mike Neushul (Neushul Mariculture) explored the possibility of farming giant kelp (*Macrocystis pyrifera*) for the production of biogas. North and his colleagues operated a "hemidome" in-field culture unit and Neushul arranged wild plants in a farm configuration to assess the growth potential of *Macrocystis*. This project was funded by the Gas Institute of America and coordinated by General Electric. The research was terminated when oil prices fell, and current research focusses on areas where the price of oil remains high; for example, the Arctic.

To produce your own kelp biogas, finely chop about 1 kg (2.2 lbs) of fresh kelp and place with 15 mL (a tablespoon) of stinky, anaerobic mud (preferably from the seashore) in a small-mouthed 4 L (1 gal) glass jar. Fill the jar about three-quarters full with chlorine-free fresh water and stopper tightly with a breathing tube (allows gas to pass out but not in; available at wine making shops). Keep warm (up to 35°C/85°F) somewhere where a little smelly gas will not be disturbing. After three to five days seal a plastic bag over the breathing tube to capture the gas. Handle the gas with care, as it is explosive and may cause olfactory displeasure!

Nutrition and Cooking

Sea vegetable products include whole dried, salted or fresh/fresh frozen plants, and flaked, powered or floured dried material. The final product may be a tea, soup mix or general seasoning, or a vegetable in stir-fry or chowder. Sea vegetable condiments may be "pure" or mixed with other spices. Healthy snacks, variously flavoured kelp chips and dulse (from the red alga *Palmaria*, p. 115), are now on the market. One company in Great Britain produces a variety of "Kelp Crunchies."

The nutritional value of sea vegetables varies considerably between the red, green and brown seaweeds. The red and green seaweeds are composed of carbohydrates that are readily digestible by humans, whereas brown algal carbohydrates are not digestible. Thus what is an energy source in red and green seaweeds is fibre in brown algae.

All sea vegetables provide some vitamins and a wide array of micronutrients. Seaweeds are famous for their ability to take up and concentrate almost all known chemical elements, as shown in this table.

Element	Concentration in sea water	Concentration in dry *Macrocystis integrifolia*	Times concentrated over sea water
Iodine	0.06 ppm*	1,016 ppm	16,933
Iron	0.003	127	42,333
Zinc	0.005	23	4,600

*parts per million (1,000 ppm=0.1%)
Source: L.D. Druehl (2000), *Parks Canada Kelp Guide*, submitted.

The ability of seaweeds to concentrate various elements is influenced by the element's availability and the characteristics of the seaweed species. For example, iodine concentration of 4,500 ppm has been recorded in *Laminaria saccharina* from BC. This is 75,000 times concentration over sea water. It was in kelp that the element iodine was first discovered, by a French chemist. The major sources of this essential element are kelp and Chile saltpeter deposits.

Public interest in sea vegetables rises with our knowledge and appreciation of their nutritional value and their potential in promoting good health (see Utilization and Cultivation of Seaweeds, above), and our desire for a varied and interesting diet. Seaweeds are excellent sources of many micronutrients and some vitamins, as shown in this table:

Component	Percent of dry weight	
	Kelp	Cabbage
Protein	3.03	1.21
Lipid	0.64	0.18
Cholesterol	0.00	0.00
Digestible carbohydrate	0.00	5.37
Fibre	9.68	0.80
Calcium	0.15	0.47
Iron	0.002	0.006
Magnesium	0.107	0.015
Phosphorus	0.080	0.023
Potassium	0.050	0.246
Zinc	0.0004	0.0002
Manganese	0.0014	0.0002

Source: *Composition of Foods*, Agriculture Handbook No. 8-11, US Department of Agriculture, Human Nutrition Information Service, 1984.

As seen in the following table, *Porphyra* (nori) is an excellent source of vitamins.

Vitamin	*Porphyra* (nori)	*Ulva* (sea lettuce)	*Laminaria* (kombu)	Spinach	Cabbage
B1 (mg/100 g)	0.21	0.06	0.08	0.12	0.05
B2 (mg/100 g)	1.00	0.03	0.32	0.30	0.05
Niacin (mg/100 g)	3.00	8.00	1.80	0.30	0.02
C (mg/100 g)	20.00	10.00	11.00	100.00	44.00
B6 (mg/100 g)	1.04	no data	0.27	0.18	0.16
B12 (ug/100 g)	21.00	6.30	0.30	0.00	0.00

Source: Seibin and Teruko Arasaki (see Annotated Selected Readings).

Unfortunately, seaweeds can also concentrate chemical elements such as lead, cadmium, copper, radioactive substances and other undesirable elements. For this reason, care must be taken selecting seaweeds or any other food for human consumption or for fertilizer or animal feed. Generally, seaweeds from heavily populated centres and industrialized regions should be avoided. Also, there is no guarantee that seaweeds from "pristine" waters contain lower levels of undesirable chemicals. These elements may be leached from natural rock formations or from abandoned mine sites and ocean dumps. Responsible producers of sea vegetables check the chemical composition of their products.

Cooking with sea vegetables is a delight. They may be prepared simply or incorporated into elaborate dishes. In the suggestions and recipes that follow, the emphasis is on western-oriented cuisine, but the use of sea vegetables is also famous in Asian cooking and many excellent recipes are available (see Annotated Selected Readings, p. 170).

Sea vegetables lend themselves to the inspirational style of cooking, where the cook muses: "Hmm, what if I were to add a...?" Dried or dried and roasted sea vegetables may be crunched into a variety of dishes. If the dish is to be cooked, adding dried sea vegetables is fine. If it is not to be cooked, you may want to use roasted seaweed. Uncooked kelp may feel slimy in your mouth. The types of kelp available for cooking all vary in texture, taste and appearance, so some kelp may be more pleasing in one dish than another. On our shores, in addition to the kombu (*Laminaria*) popular in Asian cooking, *Macrocystis*, *Nereocystis* and *Alaria* are commercially available. All three are more tender and, in my opinion, tastier than kombu. Experiment with the various types to discover what you like best.

Nori (*Porphyra*) is probably one of the healthiest foods on our planet—and this is not easy for a kelpophile to say. *Porphyra* is an excellent survival food. It is rich in carbohydrates, proteins and vitamins. It is easy to find: it grows throughout the year, high on the beach, where it is accessible. And it is easy to cook. Fresh or partially dried *Porphyra* can be roasted over a campfire, much as you would toast a marshmallow. Alternatively, dried nori can be dropped into a very hot dry pan, where it will quickly pop into popcorn-like tufts.

Dried sea vegetables can be crunched into stews, chowders, boiled potatoes, and rice and bean dishes. Bite-sized pieces can be added to stir-frys and steamer vegetables. This use of seaweeds reduces or eliminates the need to salt your food, the beauty being that the salty flavour it imparts to your food comes from a variety of salts, such as sea salt, and not just sodium chloride.

Campfire cooking can benefit from kelp. Use it to wrap potatoes, corn, fish and other foods to be cooked in the hot coals. You won't need aluminum foil, and the kelp imparts a pleasant sea-salty flavour to the foods.

Some say that kelp (*Laminaria*) reduces flatulence caused by eating beans, if it is placed in the pot with the beans as they soak and cook. The kelp is then discarded before eating. It is worth a try!

The following recipes are provided by Rae Hopkins of Canadian Kelp Resources Ltd., Bamfield, BC. Welcome to her seaweed kitchen!

If you are one of the many people who would like to try sea vegetables but have no idea how to prepare them, these simple recipes, which use some of the more readily available seaweeds, are a good place to start. But don't stop here—you can use sea vegetables in almost any dish. Dried and semi-dried sea vegetables make wonderful garnishes, too. Just remember that dried kelp increases in size when it is put into water—small pieces usually "grow" to a good bite size. Have fun, experiment, and enjoy cooking with sea vegetables!

—Rae Hopkins

Kelp Flakes or Kelp Meal

Keep a shaker of kelp flakes on the table and try it in place of salt and pepper. Any dried kelp can be used, but *Alaria* is recommended.

Using scissors, cut dried kelp into thumb-sized bits. Roast on a cookie sheet in a 180°C (350°F) oven for 15–20 minutes, being careful not to let the kelp burn. When it has cooled enough to handle, grind in a food processor or with a mortar and pestle. Store in an airtight container.

This ground kelp can be used in recipes calling for kelp flakes, powder or flour. It is also very delicious sprinkled liberally on any food you might salt. Try it on cantaloupe or apple slices.

A local sea vegetable company markets a delicious kelp flake made from *Alaria*, using this jingle:

Let me your salt and pepper be,
For I'm rich in minerals from the sea!
Iodine and calcium,
Manganese, zinc and potassium,
Iron and magnesium
And others will lead you to a life
Fantasium!

Kelp Egg Noodles

Try this interesting savoury pasta for a change of pace.
125 mL (1/2 cup) kelp flakes
250 mL (1 cup) flour
20 mL (1 Tbsp + 1 tsp) cold unsalted butter
5 mL (1/8 tsp) salt
2 large egg yolks
2 large eggs

Stir flour and kelp flakes together in a large bowl. Cut in butter and salt with a pastry blender or your fingers to form fine crumbs.

Make a well in the centre of the blended mixture. Lightly beat together the eggs and yolks, and pour them into the well. Using a fork, gradually combine the flour mixture and eggs, and continue to mix until the dough comes together.

Divide the dough into quarters. Roll out one piece with a rolling pin, stretching it a little more with each roll. Between each rolling and stretching, continue to sprinkle it with flour to keep it from sticking. Repeat until the dough is paper-thin and translucent. Repeat with remaining dough. Let the dough dry on a pasta rack or dowel for about 20 minutes. Avoid over-drying.

Roll up the noodle sheets and cut to the desired width. The noodle dough can also be rolled and cut with a hand cranked pasta machine.

Makes about 500 mL (2 cups) fresh pasta. Store in airtight container in refrigerator. To serve, cook briefly in boiling water.

Seaweed Quiche

Alaria is a versatile vegetable that can be used in place of spinach in most recipes.

Crust:
60 mL (1/4 cup) kelp flakes
625–750 mL (2 1/2–3 cups) flour
5 mL (1 tsp) baking powder
5 mL (1 tsp) salt
225 g (1/2 lb) shortening
1 egg
10 mL (2 tsp) vinegar
cold water

In a mixing bowl, stir together the kelp flakes, flour, baking powder and salt. Cut in shortening. In a measuring cup, lightly beat the egg. Add the vinegar and enough cold water to make 125 mL (1/2 cup). Pour into the dry mixture and stir until just combined.

Roll out crust and place in a 23–25 cm (9–10") pie pan.

or try a whole wheat crust:
60 mL (1/4 cup) kelp flakes
375 mL (1 1/2 cups) whole wheat flour
5 mL (1 tsp) salt
375 mL (1 1/2 cups) wheat germ
125 mL (1/2 cup) margarine
125–150 mL (1/2–2/3 cup) cold water

In a mixing bowl, stir together the kelp, flour, salt and wheat germ. Cut in the margarine. Add just enough cold water to bind the dough.

Roll out crust and place in a 23–25 cm (9–10") pie pan.

Filling:
125 mL (1/2 cup) dried *Alaria*, cut into strips (use scissors)
125 mL (1/2 cup) chopped green onion
250 mL (1 cup) sliced mushrooms
250 mL (1 cup) grated cheddar *or* Swiss cheese
3 eggs, beaten
250 mL (1 cup) light cream
125 mL (1/2 cup) milk
5 mL (1 tsp) dry mustard
1 mL (1/4 tsp) cayenne
30 mL (2 Tbsp) butter *or* margarine

Soak the *Alaria* strips in water while chopping and slicing the rest of the ingredients.

Discard soaking water. Stir together the *Alaria*, green onions, mushrooms and cheese, and spread in prepared pie crust. Combine eggs, cream, milk, dry mustard and cayenne. Pour over cheese mixture and dot with butter or margarine. Bake at 180°C (350°F) for approximately 45 minutes or until set. Serves 6.

Suimono (Kombu-dashi)

A delicate, seafood-flavoured clear soup.

Soak four pieces of dried kombu, each 10 cm (4") square, in about 1 L (4 cups) water for an hour. Bring slowly to a boil and then remove the kombu.

This soup can be enjoyed as is, or tofu pieces, sesame seeds and sliced green onion can be added. Serves 2.

Sushi

A lot of people who eat sushi don't realize the wrapper on cones or sliced sushi rolls is *Porphyra*, a.k.a. nori. A few easy fillings are suggested here, but the possibilities are endless!

4 sheets toasted nori
500 mL (2 cups) sticky rice
dried kombu
60 mL (1/4 cup) rice vinegar
550 mL (2 1/4 cups) water
5 mL (1 tsp) salt
22 mL (1 1/2 Tbsp) sugar
wasabe powder
pickled ginger

Fillings:
115–225 g (4–8 oz) smoked salmon
1 egg, beaten and cooked in frying pan, turning to cook both sides and cut julienne
2 avocados, thinly sliced
shiitake mushroom: if dried, soften in water and simmer in 125 mL/ 1/2 cup soaking water, 15 mL/1Tbsp soy sauce and 15 mL/1 Tbsp sugar until most of the liquid has evaporated; if fresh, sauté with butter and soy sauce to taste and cut into strips
large carrot, cut in thin rectangular pieces
cucumber, cut in slender lengthwise strips

Combine rice and stock in a pot with a tight-fitting lid. Bring to a boil over high heat. Reduce heat to low and steam 15 minutes. Remove from heat, cover and let stand 10 minutes.

Combine vinegar, salt and sugar. Sprinkle over rice and mix. Prepare a small amount of wasabe by stirring water into wasabe powder a few drops at a time until desired consistency.

On bamboo rolling mat (smooth side up) or clean kitchen towel, lay a sheet of toasted nori. Spread one-fourth of the rice, leaving margins on the two narrow opposite sides for nori to overlap when rolled. Spread a band of wasabe across the rice, then layer on three of the different fillings. Carefully lift the near side of the mat or cloth and roll fillings toward the far side. When roll is complete, press lightly to adjust shape.

Cut roll into 8 equal slices.

Serve with wasabe for dipping and pickled ginger. Makes 32 pieces.

Blancmange

An interesting marine delight that can be a pudding or pie filling.

1 L (4 cups) whole milk
375 mL (1½ cups) *Chondracanthus* or *Mazzaella*, dried, cut into strips and soaked in water to soften
250 mL (1 cup) strawberry preserves
15 mL (1 Tbsp) frozen orange juice concentrate
scant pinch of salt

Pour milk into a large bowl set over a pan of boiling water, double boiler style. Tie the cut seaweed in a square of cheesecloth and suspend in milk. Simmer at least 30 minutes, pressing and squeezing bag occasionally to release gel. Stir frequently.

Remove from heat and discard seaweed. Add strawberry preserves, juice concentrate and salt.

Pour the blancmange into individual serving dishes or a graham cracker pie crust. Refrigerate several hours before serving—overnight is best. Garnish with whipped cream and a sprig of fresh mint or grated chocolate. Serves 10.

Marnie's Layered Veggies

Brady's Beach vegetarian delight! Vary the type and quantities of veggies as you like.

2 blades kombu
1 carrot, julienne
1 medium zucchini, sliced

125–250 mL (1/2–1 cup) sliced mushrooms
125 mL (1/2 cup) sliced water chestnuts
250 mL (1 cup) broccoli florets
250 mL (1 cup) acorn squash, peeled and cut in strips
salt and pepper to taste
2 mL (1/2 tsp) chopped fresh ginger
1 clove garlic, chopped
30 mL (2 Tbsp) water

Soak kombu in water until softened. Lay strips of Kombu in bottom of ovenproof baking dish and layer other vegetables on top. Sprinkle on the salt, pepper, ginger and garlic. Sprinkle water over all. Cover and bake at 180°C (350°F) for 30–45 minutes or until veggies are cooked.

Nereo Soup

A delicious soup that takes little time to prepare.

125–250 mL (1/2–1 cup) dried bull kelp blades, cut into small pieces (use scissors)
750 mL (3 cups) vegetable *or* chicken stock
125 mL (1/2 cup) cubed tofu *or* pork
125 mL (1/2 cup) broccoli florets
60 mL (1/4 cup) soy sauce
1 egg, beaten
2 green onions, thinly sliced

Combine the kelp, stock, tofu, broccoli and soy sauce, and heat to boiling. Stir beaten egg into soup. Ladle into bowls and garnish with green onion. Serves 2–3.

Marinated Alaria

A good appetizer or salad.

250 mL (1 cup) dried *Alaria*, cut in strips (use scissors)
125 mL (1/2 cup) rice vinegar
45 mL (3 Tbsp) soy sauce
15 mL (1 Tbsp) sugar
5 mL (1 tsp) salt

Optional:
thinly sliced cucumber
leaves of butter lettuce *or* other tender green
sliced green onion

Place *Alaria* strips in a bowl and cover with warm water to rehydrate. Drain.

In a small bowl or jar, combine rice vinegar, soy sauce, sugar and salt and mix well. Pour over rehydrated *Alaria*.

Serve with wheat crackers as an appetizer.

To serve as a salad, add thinly sliced cucumbers and serve on leaves of butter lettuce. Garnish with sliced green onion.

Sesame Kelp Crackers

A crunchy treat, good with smoked oysters or sardines.

375 mL (1 1/2 cups) whole wheat flour
60 mL (1/4 cup) kelp powder *or* flakes
60 mL (1/4 cup) sesame seed
60 mL (1/4 cup) oil
approximately 125 mL (1/2 cup) water

Combine the flour, kelp, sesame seed and oil, and mix until crumbly. Add just enough water to bind together. Gather dough into a ball. With a rolling pin, roll dough on a lightly floured board to 3 mm (1/8") thick. Cut into squares. Bake on an ungreased baking sheet at 180°C (350°F) until golden, about 10 minutes. Store in an airtight container after the crackers have totally cooled. Makes 3–4 dozen.

Codium Tea (Chonggak)

Codium is eaten fresh in the Philippines and Hawaii and dried, in tea, in Korea. This recipe makes a spicy tea.

Use dried powdered *Codium* as you would use tea leaves. Steep 5–10 minutes.

Kombu Tea

This is a refreshing tea.

15 mL (1 Tbsp) dried kombu, chopped or ground
.5 mL (1/8 tsp) diced fresh ginger

Steep kombu and ginger 5–10 minutes for a refreshing kelp tea. A slice of lemon is a nice addition. Makes 2 cups of tea.

Macro Soup

A tasty, light vegetable soup.

750 mL (3 cups) kombu-dashi *or* chicken or vegetable stock
250 mL (1 cup) dried *Macrocystis* blades, cut into thin strips (use scissors)

rind of 1 lemon *or* orange, cut into thin match-like strips (avoid using
the white pulp—it is bitter)
1 mL ($^1/4$ tsp) salt (optional)
30 mL (2 Tbsp) light soy sauce

Bring stock to a simmer. Add thin strips of *Macrocystis*. After the kelp
has softened, add lemon rind, salt and soy sauce. Ladle into bowls.
Garnish with thinly sliced lemon. Serves 2–3.

Macro Snack Chips

Delicious with a cool beverage!

Macrocystis blades, dried
vegetable oil for frying

Using scissors, cut *Macrocystis* into chunky chip-sized pieces. Cover
the bottom of a heavy frying pan with oil and heat. Dip each kelp bit
very briefly into hot oil. (Burned bits taste bitter.) Drain on paper tow-
els. Delicious!

Glossary

Agar:
: a complex gelling compound found in the cell walls of some red algae. It has commercial applications.

Alga (plural algae):
: a collective term for photosynthetic organisms lacking distinct cellular differentiation and elaborate protected reproduction systems. Includes some bacteria, protists, and plants.

Algin:
: a complex emulsifying compound found associated with the cell walls of brown algae. It has commercial applications.

Amorphic:
: without defined form.

Anaerobic:
: lacking oxygen.

Annual:
: living for one year (cf. *Ephemeral, Perennial*).

Apical growth:
: growth resulting from cells located at the apex or along the margins of the plant (cf. *Intercalary growth*).

Asexual:
: a form of reproduction that does not involve sexual fusion, e.g. spores, fragments (cf. *Sexual*).

Axil:
: the upper angle created by a leaf coming off a stem (or a blade coming off a stipe).

Basal:
: at the base.

Binomial:
: the two names for a species (e.g. *Laminaria saccharina*).

Biota:
: all of the inhabitants of an area, usually the combined fauna and flora.

Calcareous:
: containing magnesium and/or calcium carbonate (lime) to the point of being brittle, bone-like.

Canopy:
: layers of foliage held above the ocean bottom, most significantly produced by the kelp *Macrocystis* and *Nereocystis*, but may be produced by lower-lying forms as well.

Carrageenan:
: a complex gelling compound found in the cell walls of some red algae. It has industrial uses.

Chlorophyll: the photosynthetic pigment found in all photo-synthetic organisms.

Chloroplast: the cell structure responsible for photosynthesis.

Coenocytic Cells: long cells, usually with many nuclei, that are incompletely isolated by cell end walls from adjoining cells.

Collenchyma: a type of plant cell that has minimal secondary thickening of the cell wall and gives the plant physical support until schlerenchyma has been produced.

Crust: a growth form in which the plant adheres tightly to the substrate (crustose).

Diffuse growth: growth resulting from cell divisions more or less evenly spread over the plant and not concentrated in particular areas (cf. *Meristem*).

Digitate: shaped like the fingers on a hand (refers to the shape of a seaweed blade).

Dioecious: (describing an individual plant) bearing either a male or female reproductive structure but not both (cf. *Monoecious*).

Diatom: a simple photosynthetic organism, usually a solitary cell that is encased in a silicon housing. Diatoms are common phytoplankton.

Diploid: (describing a cell) containing two sets of chromosomes (cf. *Haploid*).

Dorsal-ventral: having a differentiated front and back.

Electron microscope: a microscope capable of resolving very small structures, such as details of the chloroplast structure, which cannot be resolved by the light microscope.

Emulsifier: a chemical compound that suspends other compounds.

Ephemeral: short-lived, usually much less than one year (cf. *Annual, Perennial*).

Epiphyte: an organism that grows on another organism without harming it.

Filamentous: thread-like (describing plants usually one cell or a few cells wide).

Flagella: whip-like cellular appendages used to move a cell (e.g. some spores and sperm).

Flange: a flattened rim, resembling the brim of a hat.

Fucoid: a member of the order Fucales. Includes *Fucus, Sargassum* and related forms.

Fucoxanthin: a photosynthetic pigment found in brown seaweeds, which gives the seaweeds their brown colour.

Gamete: a cell capable of sexual fusion, usually eggs and sperm.

Gametophyte: the plant phase that bears one set of chromosomes (haploid) and produces eggs and sperm (gametes) (cf. *Sporophyte*).

Geniculate: jointed. In calcareous red algae, the term refers to alternating calcified and uncalcified segments.

Habit: the physical appearance of a seaweed or other organism.

Haploid: (describing a cell) containing one set of chromosomes (cf. *Diploid*).

Hapteron (plural haptera): finger-like projections of the holdfast that attach a seaweed to its substrate.

Herbivory: eating vegetative matter.

Heteromorphic: (describing a seaweed) having two plant phases with different morphologies (cf. *Isomorphic*).

High intertidal: the upper third of the intertidal zone, an area that is exposed to air more than water.

Holdfast: the organ of a seaweed by which it attaches.

Intercalary growth: growth that takes place between two parts of a plant. Growth that is concentrated in the middle of a filament, or between the stipe and blade, is referred to as an intercalary meristem (cf. *Apical growth*).

Intertidal: the region of beach extending between the highest and lowest levels reached by the tide. This region is exposed to air at least occasionally (cf. *Subtidal*).

Isomorphic: (describing a seaweed) having two plant phases with the same morphology (cf. *Heteromorphic*).

Life cycle: the events (meiosis, sexual fusion) and phases (gametophyte, sporophyte) through which an organism passes to enable it to reproduce sexually.

Lignin: a complex cell wall material found in advanced plants. It strengthens the plant.

Low intertidal: the lower third of the intertidal zone, an area that is exposed to water more than air.

Macroscopic: being large enough to see with the unaided eye (cf. *Microscopic*).

Meiosis: a type of cell division that reduces the sets of chromosomes from two (diploid) to one (haploid) (cf. *Mitosis*).

Meristem: a region on a plant where cell division and growth are prevalent, apical or intercalary meristems (cf. *Diffuse growth*).

Microscopic: being too small to see with the unaided eye (cf. *Macroscopic*).

Mid-intertidal: the middle third of the intertidal zone, an area that is exposed to air and water for approximately the same duration.

Midrib: a regular thickening along the middle of a seaweed blade.

Mitochondrion (plural mitochondria): the cellular structure responsible for the release of chemically bound energy.

Mitosis: a type of cell division that maintains the number of chromosome sets at their original level while increasing the number of cells (cf. *Meiosis*).

Mobile: capable of movement. In describing seaweeds, the term refers to movement resulting from flagellar activity on spores or gametes.

Monoecious:	(describing an individual plant) bearing both male and female structures (cf. *Dioecious*).
Morphology:	the study of the structures and forms of organisms.
Mutigen:	a substance that induces mutation.
Parasite:	an organism that lives on another organism, causing harm to the host. In the red algae, parasitism involves the host providing nutrition to its parasite.
Parenchyma:	a type of plant cell lacking secondary cell walls and retaining the ability to undergo cell division.
Peltate blade:	a blade whose stipe is attached near its centre and not to its margin.
Penultimate branches:	those branches next to the outermost branches or the most recently produced branches.
Perennial:	living for more than one year (cf. *Ephemeral, Annual*).
Phenotypic plasticity:	the ability to change form or function in response to environmental change.
Pheromone:	a chemical attractant produced by the female that directs the sperm to the egg.
Photoperiod:	the duration of light relative to darkness over a twenty-four-hour period that affects a plant, e.g. long- and short-day plants.
Photosynthesis:	the use of energy from light and the process of converting it into organic compounds necessary to growth and maintenance.
Phycobilins:	water-soluble photosynthetic pigments found in red algae and some photosynthetic bacteria.
Phycology:	the study of algae, including seaweeds.
Phycologist:	one who studies algae.
Phyllotaxy:	the arrangement of leaves down a stem (or blades down a stipe).
Phytoplankton:	microscopic photosynthetic organisms, usually suspended in the water, that drift with the water.

Polysiphonous: composed of tiers of elongated cells, stacked as bundles of straws. Found in some red algae.

Schlerenchyma: a type of plant cell with considerable secondary thickening of the cell wall that gives the plant its major physical support.

Sessile: having no stipe.

Sexual: a form of reproduction that involves the fusion of gametes (e.g. egg and sperm) (cf. *Asexual*).

Sib sperm: the sperm-produced individuals closely related to the recipient female.

Sorus (pl. sori): a patch of cells that have undergone meiosis, producing spores.

Splash zone: the region just above the highest tide level that receives sea water spray.

Sporophyll: a specialized blade bearing spore patches (cf. *Vegetative blade*).

Spores: small, usually single-celled structures that are produced by an organism for reproduction. Spores produced by meiosis participate in the organism's life cycle and spores produced by mitosis usually are a form of asexual reproduction. Many spores are motile.

Sporophyte: the plant phase that bears two sets of chromosomes (diploid) and usually produces spores by meiosis (fucoids are an exception).

Spray pool: a pool of water located above the highest tide level but influenced by sea water spray.

Stipe: a stem-like structure that bears the blades of seaweeds.

Storage products: energy-rich compounds produced and stored by seaweeds for later use in supporting plant maintenance, growth and/or reproduction. Starches, sugars, oils and fats may be stored.

Subtidal: the region of beach below the lowest tide level, never exposed to air (cf. *Intertidal*).

Tetrasporophyte: a diploid plant that produces spores in clusters of four by meiosis. Usually refers to a free-living sporophyte in the life cycle of many red algae.

Translocation:	the active conducting of compounds through special cells at rates that cannot be accounted for by diffusion alone.
UV:	ultraviolet light, a harmful portion of sunlight.
Vegetative blade:	a blade that is primarily responsible for photosynthesis and growth of an organism. It may also be the site of spore production (cf. *Sporophyll*).
Xylem:	specialized cells found in advanced plants that conduct water and nutrients from the roots to the above-ground plant tissues.
Zooplankton:	small animals, usually suspended in water, that drift with the water.
Zygote:	the fertilized egg cell, the first cell of the new diploid generation.

Annotated Selected Readings

The following references explore various aspects of seaweeds, their biology, their uses and those who study them. Some of the references are very technical and some are not, but even the most technical of treatments is informative for the lay person or student. These books and journals are available in university libraries and some public libraries.

Books for the identification of local seaweeds

Abbott, I.A. and G.J. Hollenberg (1976). *Marine Algae of California.* Stanford, CA: Stanford University Press. 827 pages.

This comprehensive treatment of California seaweeds had its origin in G.M. Smith's (1944) *Marine Algae of the Monterey Peninsula, California.* It has been expanded, thanks to numerous contemporary local studies. Many of the 669 species detailed in this monograph are found distributed from Alaska to Mexico. Each of them is illustrated by a lovely line drawing. An extensive glossary assists the reader in understanding the terminology used to describe seaweeds. Professor George F. Papenfuss contributed an exhaustive sketch of the historical landmarks in the development of our understanding of Pacific North American seaweeds. Many of the scientific names used in this book are out of date, but it remains the only illustrated comprehensive treatment of our seaweeds.

Gabrielson, P.W., T.B. Widdowson, S.C. Lindstrom, M.W. Hawkes and R.F. Scagel (2000). *Keys to the Benthic Marine Algae and Seagrasses of British Columbia, Southeast Alaska, Washington and Oregon.* Phycological Contribution no. 5. University of BC, Department of Botany. 187 pages.

This comprehensive guide to the identification of seaweeds considers approximately 639 forms (mostly species). Many of them are also found in California, particularly north of Point Conception. Older names of the various seaweeds are presented along with their new identities. Thus, many out-of-date seaweed names in the Abbott and Hollenberg (1976) treatment of California seaweed flora can be corrected by referring to this work. A glossary is provided to assist in using the identification keys. This guide is a companion volume to the Scagel and colleagues (1989) seaweed synopsis (see below).

O'Clair, R.M. and S.C. Lindstrom (2000). *North Pacific Seaweeds*. Auke Bay, AK: Plant Press. 159 pages.

Many of the species noted in this guide are found throughout the northeast Pacific. In addition to species descriptions, there are numerous notes on their biology. Each species is illustrated by a line drawing and given a common name, even when none existed previously.

Scagel, R.F., P.W. Gabrielson, D.J. Garbary, L. Golden, M.W. Hawkes, S.C. Lindstrom, J.C. Oliveira and T.B. Widdowson (1989). *A Synopsis of the Benthic Marine Algae of British Columbia, Southeast Alaska, Washington and Oregon*. Phycological Contribution no. 3. University of BC, Department of Botany. 532 pages.

This synopsis provides the reader with the classification and distribution of all known local species—those identified in the Gabrielson and colleagues (2000) study. Perhaps more important, this book guides the reader to all publications related to local species published between 1957 and 1989. For example, 92 references to bull kelp (*Nereocystis*) are provided. The synopsis also contains short segments on various aspects of seaweed biology, history, economics and ethnobotany.

Stewart, J.G. (1991). *Marine Algae and Seagrasses of San Diego County*. La Jolla, CA: California Sea Grant College, University of California. Report no. T-CSGCP-020. 197 pages.

This study treats approximately 160 forms (mostly species), providing descriptions with some line drawings, and notes on distribution. An appendix lists name changes that have been accepted since publication of the Abbott and Hollenberg (1976) California flora.

Books on various aspects of seaweed biology

Bold, H.C. and M.J. Wynne (1985). *Introduction to the Algae: Structure and Reproduction*, 2nd ed. Englewood Cliffs, NJ: Prentice-Hall Inc. 720

pages.

An excellent introduction to the diversity of algae, including seaweeds. Emphasis is on the description of algae with notes on their biology. This is a severely referenced book, with approximately 2,900 references for follow-up reading.

Lee, R.E. (1999). *Phycology*, 3rd ed. New York: Cambridge University Press. 645 pages.

This book looks at the principles of algal biology. Lee has provided some very nice biographic sketches, including photographs, of earlier phycologists.

Lüning, K. (1990). *Seaweeds: Their Environment, Biogeography, and Ecophysiology*. New York: John Wiley & Sons. 527 pages.

This book is a great synthesis of what is understood about seaweed distributions and various aspects of their response to environmental conditions. It contains more than 2,200 references.

Lobban, C.S. and P.J. Harrison (1994). *Seaweed Ecology and Physiology*. Cambridge: Cambridge University Press. 366 pages.

Lobban and Harrison provide the reader with comprehensive chapters on various aspects of seaweed interaction with their physical and biological environment. To reflect contemporary interests, they have included chapters on seaweed cultivation and pollution. The book also contains thoughtful essays by recognized experts on different types of seaweed communities: arctic, estuary, tropical reefs, etc.

Size, P. (1993). *A Biology of the Algae*, 2nd ed. Dubuque, IA: Wm. C. Brown Publishers. 259 pages.

This book provides a clear introduction to the common features of the algae and general descriptions of the various taxonomic groups.

Van den Hoek, C. (1995). *Algae, An Introduction Phycology*. Cambridge: Cambridge University Press. 623 pages.

This very up-to-date treatment of modern algal taxonomy is well supported with 1,937 references to earlier studies.

Books focussed on ecology

Carefoot, T. (1977). *Pacific Seashores, A Guide to Intertidal Ecology*. Vancouver: J.J. Douglas Ltd. 208 pages.

This richly illustrated treatment introduces the reader to a wide range of beach plants and animals in the context of their interactions with each other and their physical environment. The book was introduced to the public in a Canadian Broadcasting Corporation radio interview with Tom Carefoot (University of BC). On the same program, Elizabeth Carefoot's belly dancing career was explored. Carefoot dedicated *Pacific Seashores* to Elizabeth, "who first wakened in me a love of animals."

Foster, M.S. and D.R. Schiel (1985). *The Ecology of Giant Kelp Forests in California: A Community Profile*. US Fish and Wildlife Service. Biological Report 85(7.2). 152 pages.

Foster and Schiel offer a guided tour of the complex ecosystem associated with kelp forests. They introduce the inhabitants of the forests, as they fit into a food web (who eats whom) and the relationships of the forests to their physical environment. There are also discussions on the impact on the forests of such uses/abuses as harvesting, pollution and recreation. The text is well illustrated with numerous line drawings and a few colour photographs.

Books on history, ethnobotany and applied phycology

Garbary, D.J. and M.J. Wynne (eds.) (1996). *Prominent Phycologists of the 20th Century*. Hantsport, NS: Lancelot Press Ltd. 360 pages.

This book consists of 39 biographical sketches of our deceased phycological heroes, many written by those who knew them personally. It is very readable and has lots of nice photographs.

Lembi, C.A. and J.R. Waaland (eds.) (1988). *Algae and Human Affairs*. New York: Cambridge University Press. 590 pages.

Twenty-one chapters, written by specialists, cover such topics as algae in space, algae as food, algae and aquaculture, and the future of phycotechnology. This treatment will give the reader insight into the potential of seaweeds as benefactors of mankind.

Turner, N.J. (1995). *Food Plants of Coastal First Peoples*. Vancouver: UBC Press. 164 pages.

Turner documents historical and contemporary food uses of three seaweed genera (*Egregia, Macrocystis* and *Porphyra*) and numerous terrestrial plants. Each species is thoroughly described and artfully illustrated.

Cookbooks

Arasaki, Seibin and Teruko (1983). *Vegetables from the Sea*. Tokyo: Japan Publications Inc. 196 pages.

This book has comprehensive general information on seaweeds and many traditional Japanese recipes.

Lewallen, Eleanor and John (1996). *Sea Vegetable and Wildcrafter's Guide*. Mendocino, CA: Mendocino Sea Vegetable Company. 128 pages.

I have not seen this book but it sounds interesting. It includes gourmet recipes, tips on bathing with seaweed, a harvesting guide and the authors' personal essays.

Madlener, Judith Cooper (1977). *The Sea Vegetable Book*. New York: Clarkson N. Potter, Inc. 288 pages.

The Sea Grant Program in the United States was one of the principal sponsors of this book, written in the 1970s. It includes notes on various seaweeds, including growth characteristics and how the plants are used, and over 100 recipes.

McConnaughey, Evelyn (1985). *Sea Vegetables*. Happy Camp, CA: Naturegraph Publishers Inc. 239 pages.

This very practical harvesting guide and cookbook contains a wide range of recipes, from (H)alarious Carrots to Scrambled Tofu and Arame.

Richfield, Patricia (1994). *Japanese Vegetarian Cookbook*. London: Judith Piatkus (Publishers) Ltd. 170 pages.

There are beautiful photographs and good menu suggestions in this book. Most of the recipes are vegan, and those that are not are clearly marked.

Journals

Botanica Marina. Published by Walter de Gruyter, Berlin (www.deGruyter.de/journals/bm).

This bimonthly journal publishes original studies on all marine plants, including bacteria, fungi and flowering plants. Often the articles stress applied aspects of the plant's biology.

European Journal of Phycology. Published by Cambridge University Press, Cambridge (www.journals.cup.org).

This quarterly journal publishes original studies dealing with the basic biology of algae. Often book reviews and announcements of meetings are included.

Journal of Applied Phycology. Published by Kluwer Academic Publishers, Dordrecht, Netherlands (www.wkap.nl).

This bimonthly journal focusses on original research with commercial orientation. Company news, general information on new products and patents are published also.

Journal of Phycology. Published by Phycological Society of America, Inc. (www.allenpress.com/jphycol).

This bimonthly journal publishes original studies into all aspects of basic algal biology. A special feature is the regular mini-review articles on rapidly advancing areas of algal research.

Phycologia. Published by the International Phycological Society.

This bimonthly journal publishes original studies in all areas of basic algal research.

Proceedings of the International Seaweed Symposium. Published in Hydrobiologia by Dr. W. Junk Publishers.

This publication is a collection of talks given once every three years at international meetings. Many papers focus on applied areas of seaweed research.

List of Species with Authorities

This list contains all of the scientific names mentioned in this guide, including names of non-seaweeds. Where a seaweed has undergone a recent name change or there is a conflict as to which is the proper name, I have listed both the new and old names (e.g. *Sylvetia* was *Pelvetia*) or the names in conflict (e.g. *Laminaria groenlandica* and *L. borgardiana*), so that the reader can recognize organisms described in the older literature. When the full species name or binomial is used, the author in parentheses first described the species and the author named next is responsible for the species' present taxonomic position. For example, "*Laminaria saccharina* (Linnaeus) Lamouroux" tells us that the sugar kelp, *L. saccharina*, was first described by Linnaeus and later placed in its present taxonomic position by Lamouroux. This information is crucial in tracking species from their original description to present status. It is also useful in clarifying situations where the binomial has been misused.

Acrosiphonia coalita (Ruprecht) Scagel, Garbary, Golden & Hawkes
Ahnfeltia fastigiata (Postels & Ruprecht) Makiendo
Ahnfeltiopsis linearis (C.A. Agardh) Silva & deCew
Alaria fistulosa Postels & Ruprecht
Alaria marginata Postels & Ruprecht
Alaria nana Schrader
Alaria tenuifolia Setchell
Analipus japonicus (Harvey) Wynne
Anthopleura (sea anemone)
Antithamnionella pacifica (Harvey) Wollaston
Ascophyllum nodosum (Linnaeus) Le Jolis
Asterocolax gardneri (Setchell) Feldmann & Feldmann
Bangia
Bossiella californica (Decaisne) Silva
Botryoglossum
Bryopsis corticulans Setchell
Calliarthron tuberculosum (Postels & Ruprecht) Dawson
Callithamnion pikeanum Harvey
Ceramium californicum J.G. Agardh
Chaetomorpha
Chlorella-like
Chlorochytrium inclusum Setchell & Gardner

Chondracanthus exasperatus (Harvey & Bailey) Hughey
Chondrus
Cladophora aegagropila (Linnaeus) Rabenhorst
Cladophora columbiana Collins
Codiolum—phase of some filamentous green algae
Codium fragile (Suringar) Hariot
Codium setchellii Gardner
Coilodesme californica (Ruprecht) Kjellman
Colpomenia peregrina (Sauvageau) Hamel
Constantinea rosa-marina (Gmelin) Postels & Ruprecht
Constantinea simplex Setchell
Corallina officinalis var. *chilensis* (Decaisne) Kutzing
Corallina vancouveriensis Yendo
Costaria costata (C.A. Agardh) Saunders
Cryptopleura ruprechtiana (J.G. Agardh) Kylin
Cymathere triplicata (Postels & Ruprecht) J.G. Agardh
Cystoseira geminata C.A. Agardh
Cystoseira osmundacea (Turner) C.A.Agardh
Delesseria decipiens J.G. Agardh
Derbesia marina (Lyngbye) Solier
Desmarestia aculeata (Linnaeus) Lamouroux
Desmarestia kurilensis Yamada
Desmarestia ligulata (Lightfoot) Lamouroux
Desmarestia viridis (O.F. Müller) Lamouroux
Dictyoneuropsis reticulata (Saunders) G.M. Smith
Dictyoneurum californicum Ruprecht
Dictyota binghamiae J.G. Agardh
Dunaliella salina (Dunal) Teodoresco
Ectocarpus dimorphus Silva
Egregia menziesii (Turner) Areschoug
Eisenia arborea Areschoug
Endocladia muricata (Postels & Ruprecht) J.G. Agardh
Enteromorpha intestinalis (Linnaeus) Nees
Enteromorpha linza (Linnaeus) J.G. Agardh
Erythrophyllum delesserioides J.G. Agardh
Fucus gardneri Silva
Fucus spirilis Linnaeus
Gastroclonium subarticulatum (Turner) Kützing
Gelidium purpurascens Gardner
Gigartina
Gonimophyllum skottsbergii Setchell
Gracilaria pacifica Abbott
Gracilariophila oryzoides Setchell & Wilson
Gracilariopsis lemaneiformis (Bory) Dawson
Gymnogongrus
Halosaccion glandiforme (Gmelin) Ruprecht

Hapterophycus
Harveyella mirabilis (Reinsch) Schmitz & Reinsch
Hedophyllum sessile (C.A. Agardh) Setchell
Hesperophycus harveyanus (Deciasne) Setchell & Gardner
Hildenbrandia
Hymenena
Iridaea
Johnson-sea-linkia profunda Earle
Kornmannia leptoderma (Kjellman) Bliding
Laminaria borgardiana Postels & Ruprecht
Laminaria dentigera Kjellman
Laminaria ephemera Setchell
Laminaria farlowii Setchell
Laminaria groenlandica Rosenvinge
Laminaria longipes Bory
Laminaria saccharina (Linnaeus) Lamouroux
Laminaria setchellii Silva
Laminaria sinclairii (Harvey) Farlow, Anderson & Eaton
Laminaria yezoensis Miyabe
Laurencia
Leachiella pacifica Kugrens
Leathesia difformis (Linnaeus) Areschoug
Lessoniopsis littoralis (Tilden) Reinke
Lithothamnion phymatodeum Foslie
Lola lubrica (Setchell & Gardner) Hamel & Hamel
Macrocystis integrifolia Bory
Macrocystis pyrifera (Linnaeus) C.A. Agardh
Mastocarpus papillatus (C.A. Agardh) Kutzing
Mazzaella cornucopiae (Postels & Ruprecht) Hommersand
Mazzaella linearis (Setchell & Gardner) Fredericq
Mazzaella splendens (Setchell & Gardner) Fredericq
Melanosiphon intestinalis (Saunders) Wynne
Melobesia marginata Setchell & Foslie
Melobesia mediocris (Foslie) Setchell & Mason
Mesophyllum conchatum (Setchell & Foslie) Adey
Microcladia borealis Ruprecht
Microcladia coulteri Harvey
Monostroma
Myelophycus
Nemalion helminthoides (Velley) Batters
Neorhodomela larix (Turner) Masuda
Nereocystis luetkeana (Mertens) Postels & Ruprecht
Odonthalia floccosa (Esper) Falkenberg
Opuntia (a cactus)
Opuntiella californica (Farlow) Kylin
Osmunda (a fern)

Osmundea spectabilis (Postels & Ruprecht) Nam
Pachydictyon coriaceum (Holmes) Okamura
Palmaria mollis (Setchell & Gardner) van der Meer & Bird
Pelvetia
Pelvetiopsis limitata Gardner
Petalonia fascia (O.F. Müller) Kuntze
Phaeostrophion irregulare Setchell & Gardner
Phyllospadix scouleri W.J. Hooker
Phyllospadix serrulatus Ruprecht
Phyllospadix torreyi Watson
Pleurophycus gardneri Setchell & Saunders
Plocamiocolax pulvinata Setchell
Plocamium cartilagineum (Linnaeus) Dixon
Polyneura latissima (Harvey) Kylin
Polysiphonia hendryi (Kylin) Hollenberg
Porphyra perforata J.G. Agardh
Postelsia palmaeformis Ruprecht
Prasiola meridonalis Setchell & Gardner
Prionitis lanceolata (Harvey) Harvey
Pseudolithophyllum muricatum (Foslie) Steneck & Paine
Pseudolithophyllum neofarlowii (Setchell & Mason) Abey
Pterosiphonia bipinnata (Postels & Ruprecht) Falkenberg
Pterygophora californica Ruprecht
Punctaria hesperia Setchell & Gardner
Ralfsia californica sensu Setchell & Gardner
Ralfsia fungiformis (Gunnerus) Setchell & Gardner
Ralfsia pacifica Hollenberg
Rhizoclonium
Rhodymeniocolax botryides Setchell
Sargassum muticum (Yendo) Fensholt
Saundersella simplex (Saunders) Kylin
Schizymenia pacifica (Kylin) Kylin
Scytosiphon simplicissimus (Clemente) Cremades
Silvetia compressia (J. Agardh) Serrão, Cho, Boo, et Brawley
Smithoria naiadium (Anderson) Hollenberg
Soranthera ulvoidea Postels & Ruprecht
Spacelaria racemosa Greville
Tetraselmis maculata Butcher
Ulothrix flacca (Dillwyn) Thruet
Ulva fenestrata Postels & Ruprecht
Ulvaria obscura (Kützing) Gayral
Urospora penicilliformis (Roth) Areschoug
Zostera japonica Ascherson & Graebner
Zostera marina Linnaeus

Index

A page number appearing in **bold face type** indicates a photograph.

Thumbnail Identification Guide

To identify seaweeds in the field, read down the following list until you find a description that fits your specimen. Turn to the page listed and compare your specimen with the description and illustration in the text. If a photo reference is shown for the species, go to the Colour Guide and compare your specimen with the photograph. If the text and/or photo does not fit your specimen, continue down the list to narrow your search.

Green Seaweeds and Seagrasses

Plants light to dark green

Brown Seaweeds

Plants yellowish to light brown to dark brown

Red Seaweeds

Plants yellowish to red to pink to purple
Plants hardened or calcareous, pink to purple